AF301433

Nitrogen Donor Ligands in the Coordination Chemistry of the Rare Earth and Alkaline Earth Metals

Synthesis - Structures - Catalysis

Stickstoff-Donorliganden in der Koordinationschemie der Seltenerd- und Erdalkalimetalle

Synthese - Strukturen - Katalyse

Zur Erlangung des akademischen Grades eines

DOKTORS DER NATURWISSENSCHAFTEN

(Dr. rer. nat.)

der Fakultät für Chemie und Biowissenschaften

des Karlsruher Instituts für Technologie (KIT) - Universitätsbereich

vorgelegte

DISSERTATION

von

Dipl.-Chem. Jelena Jenter

aus

Berlin

Dekan: Prof. Dr. S. Bräse

Referent: Prof. Dr. P. W. Roesky

Korreferent: Prof. Dr. A. K. Powell

Tag der mündlichen Prüfung: 22.04.2010

Bibliografische Information der Deutschen Nationalbibliothek

Die Deutsche Nationalbibliothek verzeichnet diese Publikation in der Deutschen
Nationalbibliografie; detaillierte bibliografische Daten sind im Internet über
http://dnb.d-nb.de abrufbar.

1. Aufl. - Göttingen : Cuvillier, 2010
   Zugl.: Karlsruhe, Univ., Diss., 2010

   978-3-86955-342-9

Die vorliegende Arbeit wurde von 2007 bis 2010 unter Anleitung von Prof. Dr. Peter W.
Roesky an der Freien Universität Berlin (FU) und am Karlsruher Institut für Technologie
(KIT) - Universitätsbereich angefertigt.

© CUVILLIER VERLAG, Göttingen 2010
   Nonnenstieg 8, 37075 Göttingen
   Telefon: 0551-54724-0
   Telefax: 0551-54724-21
   www.cuvillier.de

1. Auflage, 2010
Gedruckt auf säurefreiem Papier

   978-3-86955-342-9

... "Zähmen, das ist eine in Vergessenheit geratene Sache", sagte der Fuchs. "Es bedeutet: sich vertraut machen."...

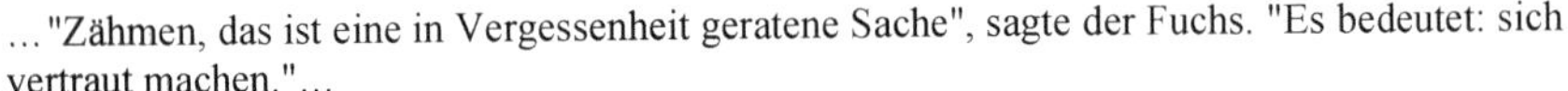

... "Hier mein Geheimnis. Es ist ganz einfach: man sieht nur mit dem Herzen gut. Das Wesentliche ist für die Augen unsichtbar."...

... "Die Menschen haben diese Wahrheit vergessen", sagte der Fuchs. "Aber du darfst sie nicht vergessen. Du bist zeitlebens für das verantwortlich, was du dir vertraut gemacht hast..."

Aus „Der kleine Prinz" von Antoine de Saint-Exupéry

Für Papa

*02.03.1948 - †10.08.2008

Table of Contents

1.   Introduction

1.1   Rare earth elements

The family of rare earth elements includes scandium, yttrium and the lanthanides. The lanthanides are the 15 elements (La-Lu), which are set apart together with the actinides from the other elements in the periodic table. Throughout the presented work, the general symbol Ln will be used to refer to all rare earth metals. The name "rare earth", which describes the metal oxides, derives from the historically late discovery of these elements. Although this term implies that these metals are less abundant than the other elements, in fact, even thulium, which is the least common lanthanide except for the radioactive promethium is more abundant in the Earth's crust than silver, mercury and the precious metals.[1] Rare earth metal compounds found versatile applications as laser materials, luminescent substances (for example in colour television sets) and permanent magnets.[2, 3] In addition, they are famous for their high activity in homogeneous catalysis as low-cost, low-toxicity and Lewis acidic catalysts.[4-8] Differently from the transition metals, the f-elements arise from the successive occupation of the f-orbitals with electrons. Regarding the lanthanides, the inert 4f shell lies deeply inside the ion and is well shielded from the 5s and 5p orbitals. This results in the lack of orbital interaction and back-bonding capability, which are important in transition metal chemistry. For this reason, lanthanides were assumed to have uninteresting chemistry. Because of their high electropositive character and the resulting tendency to form mainly ionic compounds, they were considered to be the trivalent extensions of the alkali and alkaline earth metals. The most stable oxidation state for all of the rare earth metals is +3. For a very long time, the only non-trivalent states considered accessible in coordination chemistry were Ce^{4+}, Eu^{2+}, Yb^{2+} and Sm^{2+}.[9, 10] In 2002 Evans reported molecular structures of $LnI_2(solv)$ (Ln = Tm^{2+}, Dy^{2+} and Nd^{2+}).[11, 12] Recently, organometallic complexes of divalent lanthanum were prepared.[13, 14] Due to their stability, Eu^{2+}, Yb^{2+}, and Sm^{2+} are still the most common divalent lanthanides. An important feature of the 4f-elements is called the lanthanide contraction which arises from the poor ability of f electrons to shield the other valence electrons from the nuclear charge.[2, 3] Consequently, the effective nuclear charge increases with the atomic number and the atoms/ions of the lanthanides contract. Table 1 shows the ionic radii of nine-coordinate trivalent ions of yttrium and the lanthanides.[15]

Table 1 Ionic radii for nine-coordinate trivalent rare earth metal ions.

Ln	Ionic radius [Å]	Ln	Ionic radius [Å]
La	1.216	Tb	1.095
Ce	1.196	Dy	1.083
Pr	1.179	Y	1.075
Nd	1.163	Ho	1.072
Pm	1.144	Er	1.062
Sm	1.132	Tm	1.052
Eu	1.120	Yb	1.042
Gd	1.107	Lu	1.032

Except for the late actinides, there is no other series of so many elements that exhibits similar reactivity but different ionic radii. When lanthanide complexes are applied in homogeneous catalysis, this property affords the opportunity to control catalytic reactions not only by varying the ligands, but also by altering the center metal. In comparison to transition metal ions, lanthanide ions Ln^{3+} exhibit large ionic radii and thus tend to high coordination numbers. They are strong Lewis acids, highly oxophilic and consequently very reactive towards moisture and air. The coordination chemistry of rare earth metals was dominated by cyclopentadienyl complexes for a long time. Wilkinson's pioneering work showed that there is no 18-electron rule for the cyclopentadienyl lanthanide complexes $(C_5H_5)_3Ln$.[16, 17] The triscyclopentadienyl compounds were synthesized for all of the lanthanides except for the radioactive promethium with electrons counts ranging from 18 for $(C_5H_5)_3La$ to 32 for $(C_5H_5)_3Lu$. Similarly to alkali metal or alkaline earth metal cyclopentadienyl reagents, the ionic nature of the lanthanide compounds was confirmed by the reaction with iron halides to give ferrocene. The preparation of cyclopentadienyl lanthanide chlorides[18] and dichlorides[19] gave new perspectives concerning the development of organolanthanide chemistry. Subsequently, numerous cyclopentadienyl rare earth metal complexes were synthesized[20-29] and are still of great interest in homogenous catalysis.[4-6, 28, 30-33] During the last two decades, cyclopentadienyl ligands were successfully replaced by new ligand systems to form the so-called post-metallocenes.[26, 27, 34-37] These systems are comprehensively described in the literature and some of them will be presented in the following sections.

As already mentioned, lanthanide complexes are often compared to alkaline earth metal compounds, since the bonding and structures in both species are based on electrostatic and steric factors. Regarding the divalent lanthanides, it is remarkable that the ionic radii are similar to those of the heavier alkaline earth metal ions (for six-coordinate ions: Ca^{2+} 1.00 Å, Yb^{2+} 1.02 Å, Sr^{2+} 1.18 Å, Eu^{2+} 1.17 Å, Ba^{2+} 1.35 Å).[15, 38] Therefore, it is not astounding that the coordination chemistry of these metals is related. This theory is verified by crystal structures of Ca and Yb(II) analogues, which are mostly isomorphous and have similar cell constants.[39] In addition to metallocene derivatives,[40, 41] aminotroponiminato[42] and bis(phosphinimino)methanide complexes[43] of divalent lanthanides and the heavier alkaline earth metals were prepared and all of them showed the expected similarities.

In the last decades, a great progress was achieved in the coordination chemistry of the rare earth metals. Against the initially manifested assumptions, rare earth metal compounds turned out to be widely applicable and the search for new ligand systems for creating catalytically active post-metallocenes of the rare earth metals is still in progress.

1.2 Hydroamination

Nitrogen-containing compounds, such as amines and imines, have been largely applied in the industries such as bulk chemicals, fine chemicals and pharmaceuticals.[44] Hydroamination, the addition of an amine to an unsaturated C-C bond, offers an ideal access to nitrogen-containing molecules in one step and has been intensely studied for the last decades.[33, 44-60] The direct 2+2 cycloaddition of N-H bonds to unsaturated C-C bond is orbital-forbidden. However, new reaction pathways can be achieved by using catalysts and the direct addition can be avoided. Intermolecular and intramolecular hydroamination reactions of non-activated alkenes and alkynes have been explored. As shown in Scheme 1, the intermolecular hydroamination of alkenes and alkynes can produce the Markovnikov and the anti-Markovnikov products as different regioisomers.[47, 59] In intramolecular hydroamination reactions primary and secondary aminoalkenes and aminoalkynes were used (Scheme 2). Secondary amines are more nucleophilic and thus undergo the hydroamination more facile. In general, intramolecular reactions are faster than intermolecular reactions and can be affected by varying the substituents in β-position of the amine. By using bulkier substituents, the substrate is bent into a conformation, in which the formation of the ring is facilitated; this phenomenon is established as the Thorpe-Ingold effect.[61-64]

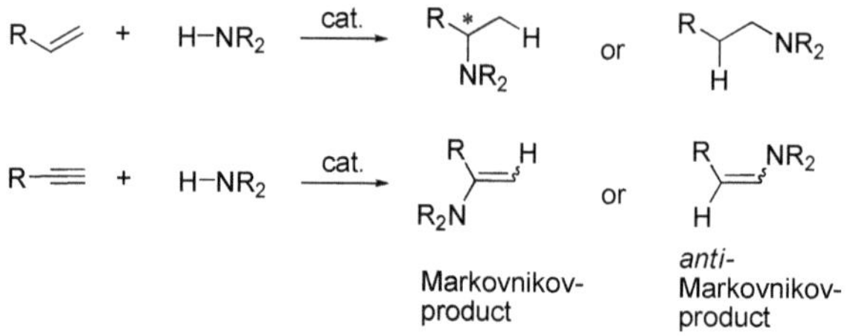

Scheme 1 Intermolecular hydroamination of alkenes and alkynes.

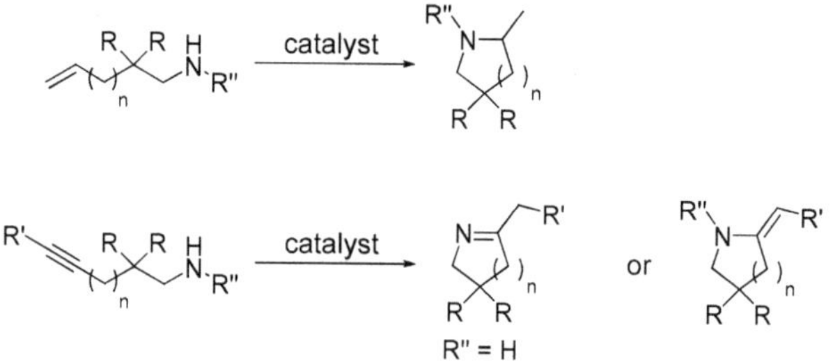

Scheme 2 Intramolecular hydroamination of aminoalkenes and aminoalkynes.

Hydroamination has an enormous reaction scope and can be catalyzed by acids, bases and organometallic compounds.[59] Brønsted acids have been used as heterogeneous[44] and homogenous[44, 59, 65] catalysts, but for latter the reaction scope is limited to amines with low basicity; otherwise the formation of ammonium salts is favoured over the protonation of the C-C double or triple bond, which is the initial step of the catalytic reaction.[59] Base catalyzed hydroamination reactions were mainly explored for alkali metals and their amides.[44, 45, 66] These basic compounds are able to deprotonate amines to give more nucleophilic metal amides which can be added to alkenes. The activation energy of these reactions is high, which is mainly caused by weak coordinative interactions between non-functionalized alkenes and alkali metal ions, and unfavorable interactions between the N lone pair and the π-system of the unsaturated C-C bond. Complexes of highly electropositive metals, such as alkaline earth and rare earth elements, show the same activation of the amine in the initial step of the catalysis and can also be classified as basic catalysts. They will be discussed more detailed later in this section. Catalytic hydroaminations were investigated for a wide range of complexes of various metals, such as alkali[45, 48, 57, 66, 67] and alkaline earth metals,[44, 48, 59] early transition metals,[44, 52, 66] late transition metals,[44, 50, 52, 54, 56, 57, 66] lanthanides and actinides.[33, 44, 50-52, 54, 55, 57-59] Late transition metals, such as palladium,[50, 52, 54, 57, 66] platinum,[52, 66] copper[52] or gold[56] tend to coordinate π-systems of unsaturated C-C bonds stronger than nitrogen donors; as a result, most catalytic cycles of hydroamination with late transition metal complexes start with the activation of the C-C unsaturated species. An exception is the activation of the amine by oxidative addition to a coordinatively unsaturated late transition metal, such as palladium, platinum or copper in low oxidation states.[59] In general, late transition metals are less sensitive to air and more tolerant of polar functional groups. On the other hand, since they are less reactive, higher temperatures and catalyst loadings are necessary. Furthermore, many efficient late transition metal catalysts contain expensive or toxic metals.[66] Group IV element complexes have been proven to be more active in hydroamination reactions of aminoalkenes and aminoalkynes than late transition metal catalysts and the reactions can proceed at lower temperatures with lower catalyst loadings.[47, 49, 50, 54, 55, 60] In addition, rare earth metal complexes are highly active catalysts for hydroamination reactions.[33, 44, 50-52, 54, 55, 57-59] Both rare earth metals and group IV metals are highly oxophilic. Thus the functional group tolerance is low and the reaction scope is limited to substrates without polar functional groups. Besides those similarities, the hydroamination reactions catalyzed by these two species exhibit different reaction mechanisms. By using group IV element complexes in hydroamination reactions, the catalytically active species is a metal imide derived from the double deprotonation of an amine. By addition of the metal imide to alkenes or alkynes an azametallacyclobutene is formed. Rare earth metal complexes deprotonate the amine only once and form a metal amide. By addition of the metal amide to alkenes or alkynes a metallacycle is formed only as transition state. A disadvantage of rare earth metal catalysts is the low reactivity in intermolecular hydroamination reactions. While intramolecular hydroamination reactions were catalyzed very efficiently, intermolecular reactions proceeded slower and with limited scope.[68] The low reactivity in intermolecular reactions resulted primarily from the competition between strongly coordinating amines and weakly coordinating alkenes or alkynes to the metal center. For this reason, the presented work focuses on the intramolecular hydroamination which has been studied intensively for the last decades.[33, 58, 59, 68-70] Scheme 3 shows the simplified mechanism generally accepted for the intramolecular hydroamination of terminal aminoalkenes.[59] As mentioned above, the amine is deprotonated by alkyl or basic amide ligands and the resulting amide is coordinated to the metal center in the initial step of the reaction (catalyst activation). The first step of the catalytic cycle is the alkene insertion into the rare earth metal amido bond *via* a cyclic transition state, in which both the amido and the alkene function are coordinated to the metal center. With the protonolysis of the cyclic amine by the substrate, the product is formed and the catalyst is regenerated to complete the catalytic cycle.

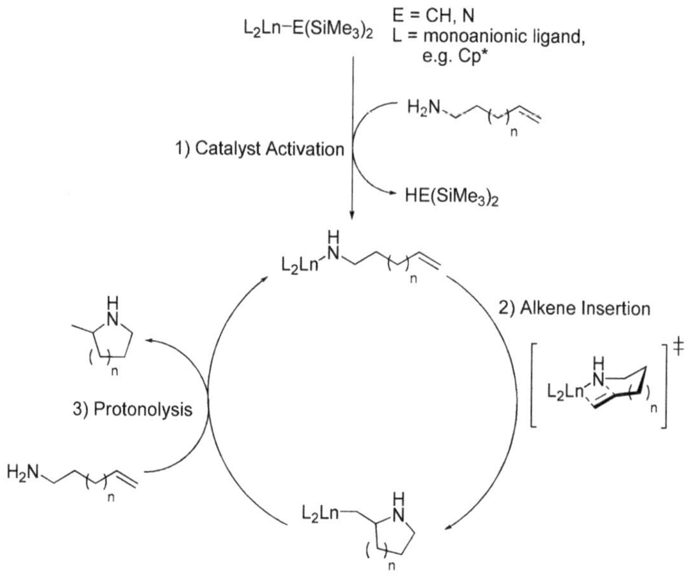

Scheme 3 The simplified mechanism of intramolecular hydroamination of aminoalkenes catalyzed by rare earth metal complexes.

The mechanism of hydroamination of aminoalkynes catalyzed by rare earth metal complexes is similar to that of aminoalkenes, but the reaction proceeds significantly faster.[70] While hydroaminations of terminal and internal alkynes exhibit comparable rates, internal aminoalkenes react remarkably slower than terminal aminoalkenes.[33, 59] Various rare earth metal complexes have been developed and investigated for hydroamination reactions of aminoalkenes and aminoalkynes. In general, characteristic features of a catalytically active rare earth metal complex are spectator ligands which stay attached to the metal center during the catalytic cycle and leaving groups which are replaced by the substrate in the initial step of the catalysis. Hydrides, alkyls, phosphides or basic amides are used as leaving groups and can be protonated and consequently replaced by the amine. Initially, cyclopentadienyl derivatives, which have been the most established ligands in rare earth metal chemistry, were used as spectator ligands.[69-71] In addition, chiral ansa-metallocenes showed high stereoselectivities in asymmetric hydroamination/cyclization reactions.[54, 72] Besides cyclopentadienyl ligands which are still of great interest for catalyst development,[33, 58] new spectator ligands were introduced to rare earth metal chemistry and various complexes with different leaving groups were synthesized. The post-metallocenes exhibit significantly more diverse steric and electronic properties and were proven to be suitable catalysts for hydroamination reactions.[73-96] The first non-cyclopentadienyl catalysts for the hydroamination/cyclization reaction contain an aminotroponiminato ligand and were reported in 1998 by our group.[90] Subsequently, a large scale of non-cyclopentadienyl rare earth metal complexes was explored for hydroamination reactions, including chiral catalysts for asymmetric hydroamination/cyclizations. Various bidentate ligands used in post-metallocenes include: *NN-monoanionic ligands*, such as amidinates,[73, 74] β-diketiminates,[76] bisoxazolinates,[80, 91] aminotroponiminates[75, 90] and bis(phosphinimino)methanides;[77-79] *NN-dianionic ligands*, such as diamides;[81-89] *OO-dianionic ligands*, such as biphenolates and binaphtholates.[91, 94-96] Furthermore, rare earth metal complexes containing tetradentate *NNNN-monoanionic*[93] or *-dianionic*[92] ligands were synthesized and investigated for hydroamination reactions. All the ligands mentioned above are comprehensively described in the literature and thus will not be discussed in detail here. Beside the spectator ligands and the leaving groups, the ionic radius of the center metal has a remarkable influence on the reaction rate. While the reaction rate of hydroamination of aminoalkenes increases with increasing ionic radius of the center metal, the rate of hydroamination of aminoalkynes decreases with increasing ionic radius. Of particular interest for the present work is the intramolecular hydroamination of aminoalkenes and aminoalkynes with rare earth and alkaline earth metal complexes. Although the chemistry of alkaline earth element complexes is closely related to that of the rare earth metal complexes, little hydroamination studies of alkaline earth metals were performed. In 2005, Hill *et al.* investigated the β-diketiminato calcium amide [{(DIPNC(Me))$_2$CH}-Ca{N(SiMe$_3$)$_2$}(THF)] (DIP = 2,6-diisopropylphenyl) for intramolecular hydroaminations of aminoalkenes.[97] As expected, similar reactivity to that of the rare earth metal complexes was observed, which showed high conversions and short reaction times at low temperature. The mechanism of alkaline earth metal complex catalyzed hydroamination of aminoalkenes is similar to that of rare earth metal complexes as well (Scheme 3). Recently, Hill *et al.* expanded their experimental and computational studies by using β-diketiminato amide complexes of different alkaline earth metals.[98, 99] In addition to the calcium compound, the β-diketiminato magnesium methyl complex [{(DIPNC(Me))$_2$CH}Mg(Me)(THF)] was synthesized; but a lower catalytic activity was observed. By using the calcium species, competitive alkene isomerization occurred as a side reaction. The proposed mechanism is shown in Scheme 4. The problem of ligand redistribution occurred for the heavier homologues strontium and barium. Alkaline earth metal compounds LMX (L = monoanionic ligand, M = alkaline earth metal, X = leaving group) tend to undergo Schlenk-type or/and irreversible ligand redistribution reactions to form L_2M and MX_2. The β-diketiminato ligand was able to stabilize the magnesium and calcium complexes kinetically, but for strontium and barium the homoleptic complexes were formed as side-products. Computational studies predict the β-diketiminato strontium amide to be the most active catalyst, followed by the analogous calcium compound.

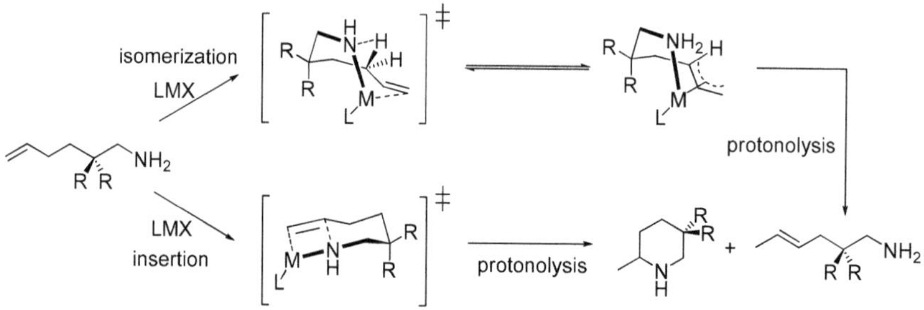

Scheme 4 Proposed reaction mechanism for the formation of the products obtained from the intramolecular hydroamination of 1-Amino-5-hexenes catalyzed by β-diketiminato calcium bis(trimetylsilyl)amide.[99]

Very recently, Hill *et al.* synthesized $[\{H_2B(Im^{tBu})_2\}M\{N(SiMe_3)_2\}(THF)_n]$ (M = Ca, n = 1; M = Sr, n = 2, Im^{tBu} = 1-*tert*-butylimidazole) as catalytically active species for the intramolecular hydroamination of aminoalkenes.[100] Different to the β-diketiminato alkaline earth metal complexes, no conclusive evidence for Schlenk-type ligand redistribution was observed for the strontium compound. On the contrary, the strontium derivative were proven to be the most active catalyst, which is consistent with the computational studies of the β-diketiminato alkaline earth metal amides.[98] The catalytic activity of the bis(imidazolin-2-ylidene-1-yl)borate calcium amide is similar to that observed for the analogous β-diketiminato calcium compound.[100] In addition, no alkene isomerization occurred as side reactions by using the bis(imidazolin-2-ylidene-1-yl)borate alkaline earth metal derivatives as catalysts. Furthermore, aminotroponiminate alkaline earth metal amides were synthesized by our group and proved to be catalytically active for the intramolecular hydroamination of aminoalkenes.[42, 101, 102] The heteroleptic complexes of calcium and strontium $[\{(iPr)_2ATI\}M\{N(SiMe_3)_2\}(THF)_2]$ (M = Ca, Sr; $(iPr)_2ATI$ = *N*-isopropyl-2-(isopropylamino)troponiminate) were obtained, but only the homoleptic complex [{(*i*Pr)$_2$ATI}$_2$Ba(THF)$_2$] was formed with barium (Schlenk-redistribution). The calcium compound gave the best results in catalytic hydroamination/cyclizations. In this case, the catalytic activity of the aminotroponiminato alkaline earth metal complexes decreases with increasing ionic radius of the metal (starting from calcium), which is contrary to the general trend observed for rare-earth metal catalysts and to the results observed for the alkaline earth metal complexes containing the β-diketiminato or the bis(imidazolin-2-ylidene-1-yl)borate ligand. Regarding alkaline earth metal catalysts, the reaction rates and selectivity cannot be simply related to the ionic radius. Further investigations of the kinetic aspects of the reactions are required. In summary, only few alkaline earth metal complexes were explored towards their catalytic activity in hydroamination/cyclizations. This field of research offers many opportunities for new discoveries.

1.3 Rare earth metal borohydrides

Borohydride compounds have been applied as potential hydrogen storage materials.[103] In 1960, Zange prepared the first rare earth metal trisborohydride complexes [Ln(BH$_4$)$_3$(THF)$_n$] by the reaction of rare earth metal alkoxides with B$_2$H$_6$.[104] In the 1980s, Mirsaidov *et al.* generated a more convenient pathway by the metathesis reaction of LnCl$_3$ (Ln = La, Ce, Pr, Nd, Sm) with NaBH$_4$.[105-107] In 2000, Cendrowski-Guillaume *et al.* optimized the reaction conditions by reducing the amount of NaBH$_4$, extending the reaction time and elevating the temperature.[108] [Ln(BH$_4$)$_3$(THF)$_n$] have been proven to be valuable precursors for the synthesis of organometallic compounds in salt metathesis reactions with the corresponding alkali metal organyls.[108-111] Besides cyclopentadienyl or cyclooctatetraene derivatives,[108-114] nitrogen or oxygen donor ligands have been used to synthesize rare earth metal borohydride complexes.[115-119] Rare earth metal trisborohydrides and their derivatives are efficient catalysts for the polymerization and copolymerization of styrene,[120-122] isoprene,[112, 114, 120, 123, 124] ethylene[114, 125] and some polar monomers such as ε-caprolactone (CL),[119, 126-135] lactide,[115, 116, 119, 132, 134] methyl methacrylate (MMA)[113, 117, 136, 137] and trimethylene carbonate.[127] Metallocene as well as post-metallocene borohydrides showed high activities and the development of new catalysts is in great demand.

1.4   The bis(phosphinimino)methanide ligand

Besides attracting much attention in main-group and transition-metal chemistry, ligands that contain P-N units have become more and more popular in rare earth metal chemistry.[138-145] Bis(phosphinimino)methanides are monoanionic NN ligand systems which have been proven to be very suitable in stabilizing rare earth metal complexes. The rare earth metal complexes containing these ligands showed interesting coordination modes as well as good catalytic activities in hydroamination, hydrosilylation and the polymerization of MMA and CL.[77, 146-149] The bis(phosphinimino)methanide framework can be formally built up by replacing two carbon atoms of the well known β-diketiminato framework with two phosphorus atoms (Scheme 5). These two ligand systems exhibit a topological relationship which appears among others in similarities in donor identity and steric demands. However, bis(phosphinimino)methanide ligands show a localized negative charge on the methanide carbon atom. Therefore this carbon atom can act as a third coordination site by building a long bonding interaction with the metal center. The resulting metallacycles show as a typical structural feature the pseudo-boat conformation (Scheme 6).

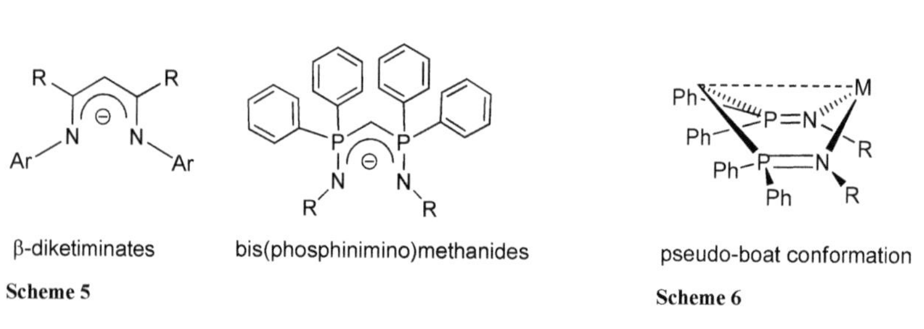

β-diketiminates   bis(phosphinimino)methanides   pseudo-boat conformation

Scheme 5   **Scheme 6**

Furthermore, in contrast to β-diketiminato ligands, the methanide group of bis(phosphinimino)methanides can be deprotonated again to give bis(phosphinimino)methandiides which act as dianionic ligand systems. As shown by Cavell *et al.*, this ligand forms a "pincer" carbene type complex with samarium, which exhibits a short bonding distance between the metal center and the methandiide carbon atom.[150, 151] Only the monoanionic bis(phosphinimino)methanide ligand systems will be presented here. Although bis(phosphinimino)methanides with different substituents (R = DIP, Mes, SiMe$_3$) were reported,[152] catalytic applications in rare earth metal chemistry have been restricted to the trimethylsilyl substituted species. According to this, the following section focuses on the {CH(PPh$_2$NSiMe$_3$)$_2$}$^-$ ligand with the corresponding rare earth metal complexes.

The bis(phosphinimino)methanide rare earth metal complexes are prepared by two general methods: salt and amine eliminations. The salt elimination method is more established. In order to avoid the problem of occlusion of solvated lithium chloride in salt metathesis reactions, the potassium bis(phosphinimino)methanide complex **1** was used as a ligand transfer reagent. The solvent free complex **1** can be easily accessed by the reaction of bis(phosphinimino)methane $CH_2(PPh_2NSiMe_3)_2$ with an excess of KH in THF (Scheme 7).[153]

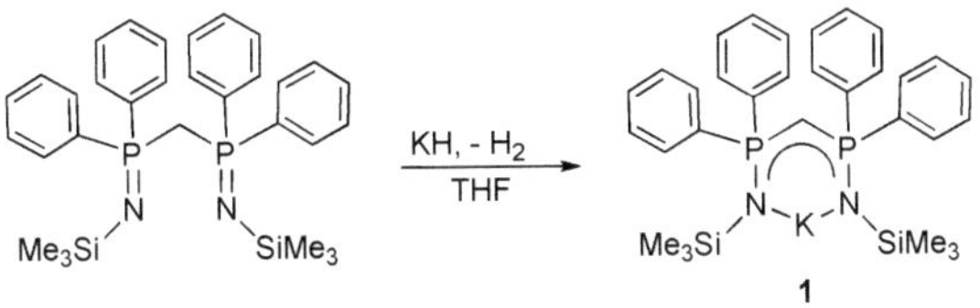

Scheme 7

The dimeric rare earth metal dichloride complexes $[\{(Me_3SiNPPh_2)_2CH\}LnCl_2]_2$ (Ln = Y (**Ia**), Sm (**Ib**), Dy (**Ic**), Er (**Id**), Yb (**Ie**), Lu (**If**)) were obtained from the salt metathesis reaction of potassium bis(phosphinimino)methanide **1** with anhydrous yttrium or lanthanide trichlorides (Scheme 8).[154] When the metal center is larger than samarium, bis(phosphinimino)methanide lanthanide dichloride complexes could not be obtained.

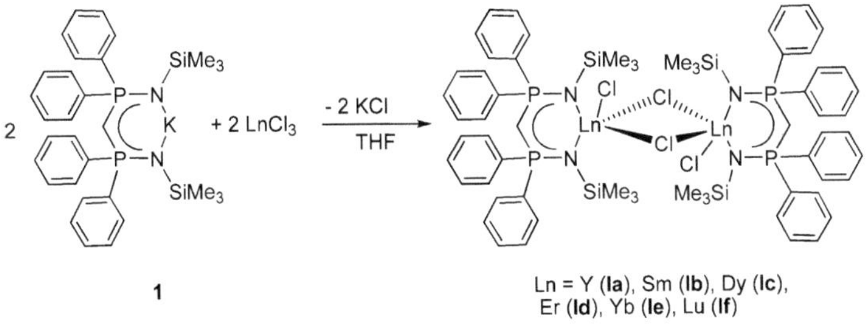

Scheme 8

In order to form new catalytically active compounds, the chloro atoms were replaced by different amines. Scheme 9 shows the resulting complexes which were investigated for catalytic reactions.

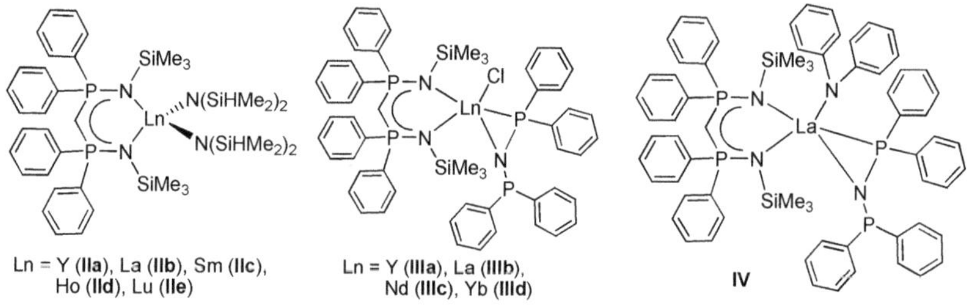

Ln = Y (**IIa**), La (**IIb**), Sm (**IIc**), Ho (**IId**), Lu (**IIe**)

Ln = Y (**IIIa**), La (**IIIb**), Nd (**IIIc**), Yb (**IIId**)

IV

Scheme 9

[{($Me_3SiNPPh_2)_2CH$}Ln{$N(SiHMe_2)_2$}$_2$] (Ln = Y (**IIa**), La (**IIb**), Sm(**IIc**), Ho (**IId**), Lu (**IIe**)) were synthesized *via* amine or salt elimination reactions.[155, 156] All complexes **IIa-IIe** are catalytically active systems for hydroamination (Section 1.2, Scheme 2), hydrosilylation (Scheme 10) and the sequential hydroamination/hydrosilylation reactions (Scheme 11).[156] Due to the rate dependence on the ionic radius of the center metal, the lanthanum complex **IIb** is the most active catalyst. All reactions afforded the products in high yields and good reaction times.

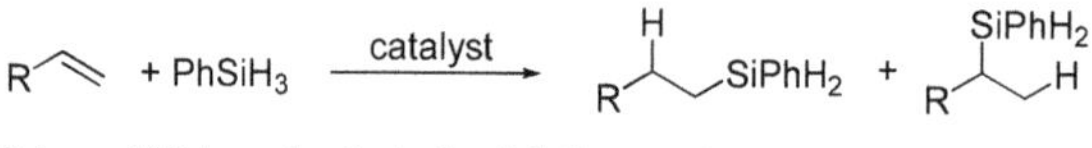

Scheme 10 Intermolecular hydrosilylation reactions

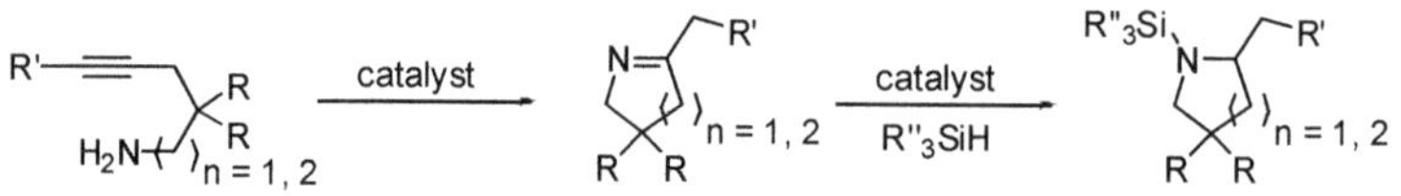

Scheme 11 The sequential hydroamination/hydrosilylation reaction

[{($Me_3SiNPPh_2)_2CH$}LnCl{$N(PPh_2)_2$}] (Ln = Y (**IIIa**), La (**IIIb**), Nd (**IIIc**), Yb (**IIId**)) (Scheme 9) were synthesized *via* salt metathesis reactions.[149] Additionally, the chloro ligand of **IIIb** was replaced by the {NPh_2}$^-$ ligand to give [{($Me_3SiNPPh_2)_2CH$}Ln{NPh_2}{$N(PPh_2)_2$}] **IV** (Scheme 9). Complexes **IIIa-IIId** and **IV** have been shown to be active catalysts for the polymerization of CL (Scheme 12) and MMA (Scheme 13).[149] In both reactions, the catalytic activity depends on the ionic radius of the metal center, and the lanthanum and the neodymium complexes **IIIb** and **IIIc** are more active initiators. The polymerization of CL afforded polycaprolactones with high polydispersities. In contrast, the polymerization of MMA catalyzed by **IIIb** led to high molecular weight polymers with good conversions and narrow polydispersities.

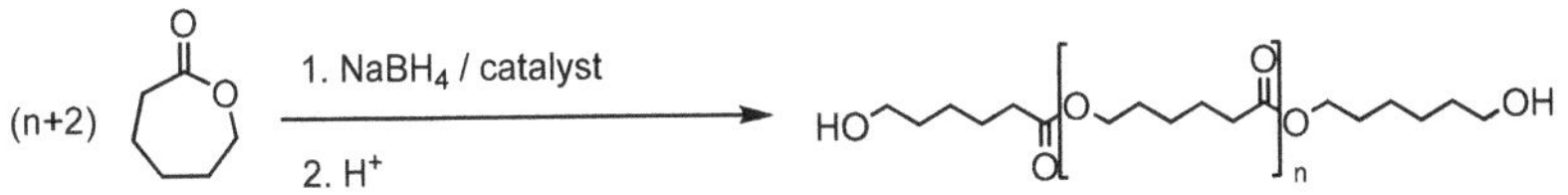

Scheme 12 Polymerization of ε-caprolactone (CL) to produce α, ω-dihydroxytelechelic polymers

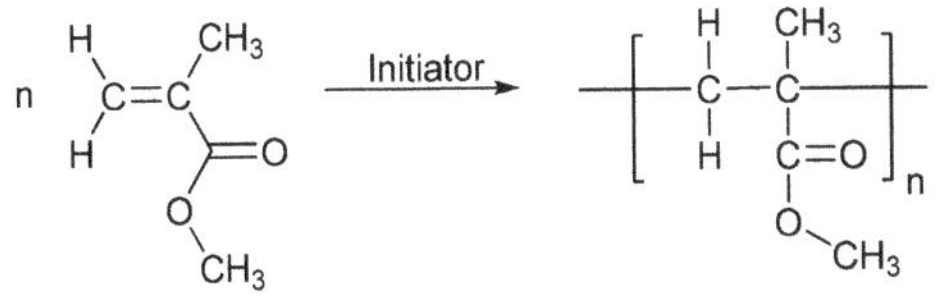

Scheme 13 Polymerization of methyl methacrylate (MMA)

In conclusion, the bis(phosphinimino)methanide $\{CH(PPh_2NSiMe_3)_2\}^-$ has been proven to be a suitable ligand in rare earth metal chemistry to form post-metallocenes. Due to the successful application in homogeneous catalysis, it is of great interest to continue varying the ligands of bis(phosphinimino)methanide rare earth metal complexes in order to form novel active catalysts.

1.5   The bis(imino)pyrrolyl ligand

2,5-Bis(*N*-aryliminomethyl)pyrrolyl compounds are monoanionic ligands which can be easily accessed by deprotonation of the corresponding bis(imino)pyrroles.[157] Bis(imino)pyrroles can be prepared by condensation of 2,5-diformylpyrrole with the analogous anilines.[158] 2,5-bis(*N*-aryliminomethyl)pyrrolyl ligands contain a pyrrolyl unit with two imino groups. Thus this ligand system exhibits three possible coordination sides. As shown in Scheme 14, bis(imino)pyrrolyl compounds can act as bidentate or tridentate chelating ligands to form **A** or **B** by coordination to a metal center.

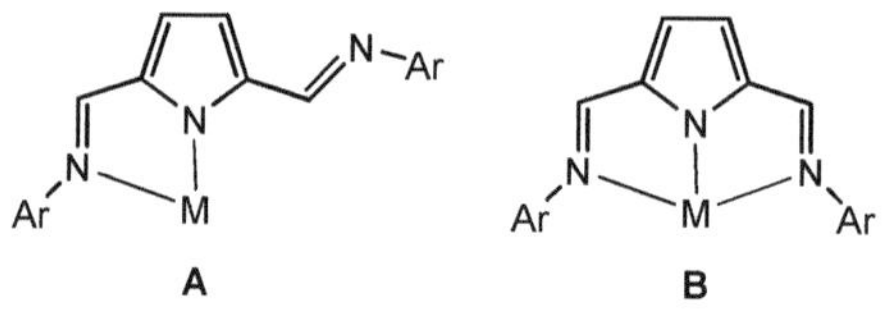

A   **B**

Scheme 14

The coordination mode of the resulting complexes depends on the character of the aryl-substituents of the ligand, the number of ligands coordinated, the ionic radius of the metal center and the bulkiness of additional ligands coordinated to the metal atom. In this context, 2,5-bis(*N*-aryliminomethyl)pyrrolyl ligands were already introduced into aluminum,[159] group 4,[160, 161] chromium,[162] iron,[160] cobalt,[160] and nickel chemistry.[160] Some of them have been used as catalysts for the polymerization of ethylene[159-162] and the oligomerization of ethylene and propylene to linear and branched products.[160] In 2001, Mashima *et al.* introduced 2,5-bis(*N*-aryliminomethyl)pyrrolyl compounds into rare earth metal chemistry by forming yttrium complexes containing various aryl substituted ligands.[158] The complexes were synthesized *via* amine elimination reactions of the corresponding pyrroles with [Y{N(SiMe$_3$)$_2$}$_3$]. The coordination mode and the number of the 2,5-bis(*N*-aryliminomethyl)pyrrolyl ligands introduced to a metal center can be controlled by varying the bulkiness on the aryl group of the ligands (Scheme 15).

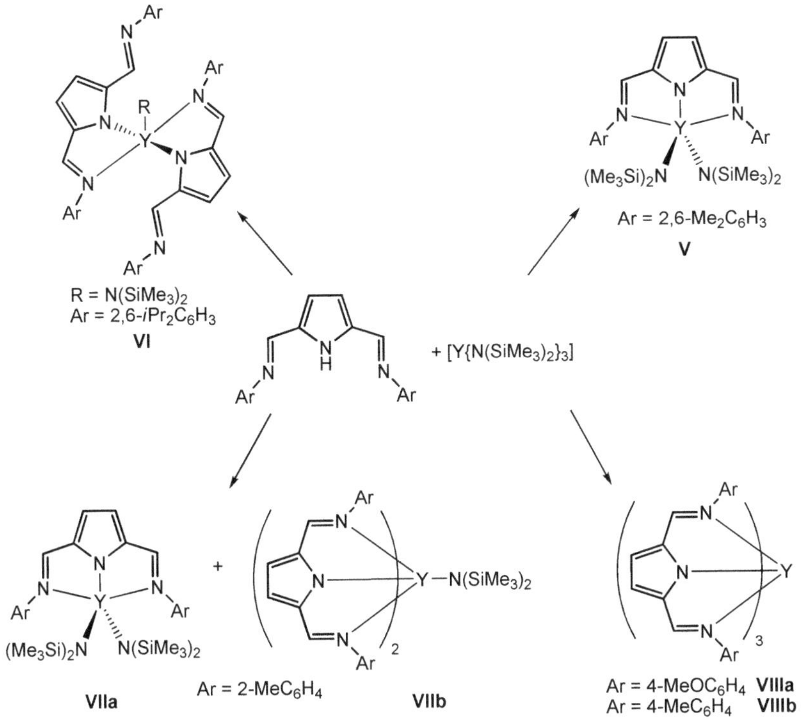

Scheme 15

The reaction of the 2,6-dimethylphenyl substituted ligand with [Y{N(SiMe₃)₂}₃] led to the mono-substituted complex **V** in which the ligand is coordinated by all three nitrogen atoms to the metal center. By using the bulkier 2,6-diisopropylphenyl substituted ligand, the bis(pyrrolyl) complex **VI** was formed, in which both pyrrolyl ligands coordinate to the metal center in a bidentate fashion. The reaction of [Y{N(SiMe₃)₂}₃] with the 2-methylphenyl substituted ligand gave a 4:1 mixture of mono(pyrrolyl) complex **VIIa** and bis(pyrrolyl) complex **VIIb**. The reaction of 4-methoxyphenyl or 4-methylphenyl substituted ligand with [Y{N(SiMe₃)₂}₃] afforded the homoleptic tris(pyrrolyl) complexes **VIIIa** and **VIIIb** in which the pyrrolyl ligands are bound in a tridentate fashion to the metal center. The catalytic activity of yttrium complexes **V**, **VI**, **VIIIa** and **VIIIb** was investigated for the polymerization of CL (Scheme 12).[158] The homoleptic pyrrolyl complexes **VIIIa** and **VIIIb** showed no activity, while complexes **V** and **VI** were active catalysts for the polymerization of CL. The polymers obtained with **VI** as catalyst exhibited narrower polydispersities compared with the polymers produced by **V**. In conclusion, the yttrium complex containing the 2,6-diisopropylphenyl substituted ligand showed the best catalytic activity and consequently, further studies in rare earth chemistry focus on this ligand system.

The rare earth metal complexes are generally synthesized by two methods: salt and amine eliminations. For salt metathesis reactions the preparation of the corresponding lithium or potassium salt of the ligand is required. The lithium complex [(DIP$_2$pyr)Li] (**9**) was synthesized by the reaction of 2,5-bis{*N*-(2,6-diisopropylphenyl)iminomethyl}pyrrole (**8**) with *n*BuLi in toluene (Scheme 16).[160, 163]

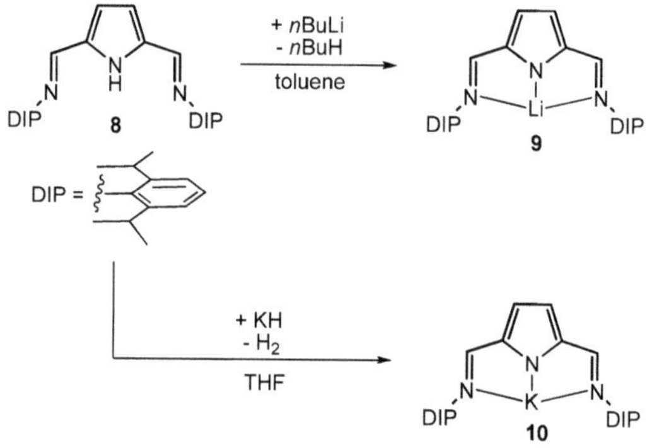

Scheme 16

A general problem in rare earth metal chemistry by using the lithium salt as a ligand transfer reagent in salt methasesis reactions is the occlusion of solvated lithium chloride to form "ate"-complexes. As shown in Scheme 17, the reaction of **9** with anhydrous yttrium trichloride afforded the "ate"-complex **IX**.[163] In order to avoid this problem, the synthesis of the potassium salt was developed by our group. The treatment of **8** with KH in hot THF afforded [(DIP$_2$pyr)K] (**10**) (Scheme 16).[163] No crystal structure has been reported for either the lithium or the potassium complex. The salt metathesis pathway for producing new rare earth metal compounds has been successfully performed for the 2,5-bis{*N*-(2,6-diisopropylphenyl)iminomethyl}pyrrolyl ligand by N. Meyer.[163, 164] The reaction of **10** with anhydrous LnCl$_3$ led to the corresponding compounds [(DIP$_2$pyr)LnCl$_2$(THF)$_2$] (Ln = Y (**Xa**), Lu (**Xb**)). As shown in Scheme 17, both complexes **IX** and **Xa** could be transformed into [(DIP$_2$pyr)Y(C$_5$H$_5$)$_2$] (**XI**) by reactions with Na(C$_5$H$_5$).[163] In the described complexes **IX**, **Xa**, **Xb** and **XI** (Scheme 17), the (DIP$_2$pyr)⁻ ligand coordinates in a tridentate fashion onto the metal center.

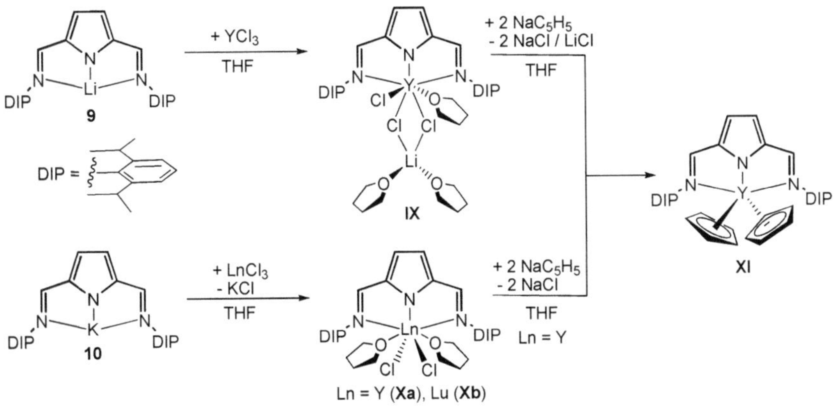

Scheme 17

The bis(imino)pyrrolyl ligand $(DIP_2pyr)^-$ was shown to be an appropriate ligand in rare earth metal chemistry. The special feature of variable coordination modes of this ligand, which has already appeared in different complexes, leads to the interesting question: what happens by using rare earth metals in different sizes? In addition, the creation of new catalytically active complexes is required.

2. Aim of the project

The general goal of the presented work was to synthesize novel post-metallocenes by using either the bis(phosphinimino)methanide $\{CH(PPh_2NSiMe_3)_2\}^-$ or the 2,5-bis{N-(2,6-diisopropylphenyl)iminomethyl}pyrrolyl ligand. The bis(phosphinimino)methanide ligand has been well established in rare earth metal chemistry (Section 1.4), while only few compounds have been prepared with the 2,5-bis{N-(2,6-diisopropylphenyl)-iminomethyl}pyrrolyl ligand (Section 1.5). The initial aim of the project was to synthesize rare earth metal borohydride complexes of both ligands, since they were expected to show high activity in polymerization reactions (Section 1.3) and to form interesting structures in the solid state. The major focus of the work was on the 2,5-bis{N-(2,6-diisopropylphenyl)-iminomethyl}pyrrolyl ligand. Initially, neodymium containing compounds, such as chlorides and borohydrides were in great demand, since they can be used as Ziegler-Natta catalysts in the polymerization of 1,3-butadiene. Furthermore, it was a great challenge to introduce the 2,5-bis{N-(2,6-diisopropylphenyl)iminomethyl}pyrrolyl ligand into the coordination chemistry of the divalent lanthanides and the alkaline earth metals. Alkaline earth metal catalysts were proven to be very active in intramolecular hydroamination reactions, but only few compounds have been hitherto explored (Section 1.2). For this reason, it was another aim to synthesize potential catalytically active alkaline earth metal complexes of this ligand to investigate them for the intramolecular hydroamination of non-activated aminoalkenes and aminoalkynes. High activities and selectivities were expected. Since the 2,5-bis{N-(2,6-diisopropylphenyl)iminomethyl}pyrrolyl ligand showed interesting coordination modes in previous studies, attractive structural motifs were expected for the new complexes.

Moreover, it was of great interest to investigate imidazolin-2-imide and cyclopentadienyl-imidazolin-2-imine rare earth metal alkyl complexes, synthesized by M. Tamm *et al.*, for their catalytic activity in the intramolecular hydroamination of non-activated aminoalkenes and aminoalkynes. Since rare earth metal complexes were proven to act as highly active catalysts in such hydroamination reactions, both compounds had the potential to show high activities and selectivities.

3.   Results and discussion

3.1   Bis(phosphinimino)methanide rare earth metal borohydrides

Aliphatic polyesters and polycarbonates, derived from heterocyclic monomers such as lactones, dilactones and carbonates, show high biocompatibility and bioresorbability.[165-173] Therefore, they are very attractive for environmental applications, such as recyclable plastic substitutes or textile derivatives, and widely used in drug-delivery systems or in tissue engineering. The ring-opening polymerization (ROP), initiated by rare earth metal derivatives, has attracted growing attention in the past two decades, since these catalysts showed high activity with limited side reactions and can control regio- and stereoselectivity of the reactions.[32, 174-176] Rare earth metal borohydrides, such as $[Ln(BH_4)_3(THF)_3]$ (Ln = La, Nd, Sm) and $[(C_5Me_5)_2Sm(BH_4)(THF)]$, have been proven to be active catalysts for the polymerization of CL.[126-128, 130, 131, 133, 134] In addition, post-metallocene borohydrides of the rare earth metals have been used as initiators for this type of polymerization.[115, 119, 129] The bis(phosphinimino)methanide ligand $\{CH(PPh_2NSiMe_3)_2\}^-$ is well established in rare earth metal chemistry by forming post-metallocenes which found versatile applications in homogeneous catalysis, such as hydroamination, hydrosilylation and polymerization reactions.[77, 146-149, 156] In the presented work, bis(phosphinimino)methanide rare earth metal borohydrides were synthesized and a number of complexes were investigated for their catalytic activity in the ROP of CL.[177]

The bis(phosphinimino)methanide rare earth metal borohydrides were successfully synthesized *via* salt metathesis reactions of $[K\{CH(PPh_2NSiMe_3)_2\}]$[153] with $[Ln(BH_4)_3(THF)_n]$ (Ln = Sc (n = 2); Ln = La, Nd, Lu (n = 3)) in THF at elevated temperature (Scheme 18). Since rare earth metal trisborohydrides could be easily accessed by the reaction of anhydrous $LnCl_3$ with $NaBH_4$,[108] the desired products $[\{(Me_3SiNPPh_2)_2CH\}Ln(BH_4)_2(THF)_n]$ (Ln = La (3), Nd (4), n = 1; Ln = Sc (5), Lu (7), n = 0) were obtained in a convenient two step synthesis. As shown in Scheme 18, a different reaction pathway was used for the preparation of $[\{(Me_3SiNPPh_2)_2CH\}Y(BH_4)_2]$ (6). The synthesis of 6 occurred in a two step synthesis as well. Initially, $[\{(Me_3SiNPPh_2)_2CH\}YCl_2]_2$ was synthesized by the reaction of $[K\{CH(PPh_2NSiMe_3)_2\}]$ with anhydrous yttrium trichloride.[154] The treatment with $NaBH_4$ *in situ* at elevated temperature afforded the desired complex 6. For the rare earth metals with larger ionic radius (Ln = La, Nd), a THF molecule is additionally coordinated to the metal center to form the complexes $[\{(Me_3SiNPPh_2)_2CH\}Ln(BH_4)_2(THF)]$ (Ln = La (3), Nd (4)). For the smaller metal ions

(Ln = Sc, Y, Lu), the solvent free complexes [{(Me$_3$SiNPPh$_2$)$_2$CH}Ln(BH$_4$)$_2$] (Ln = Sc (**5**), Y (**6**), Lu (**7**)) were obtained.

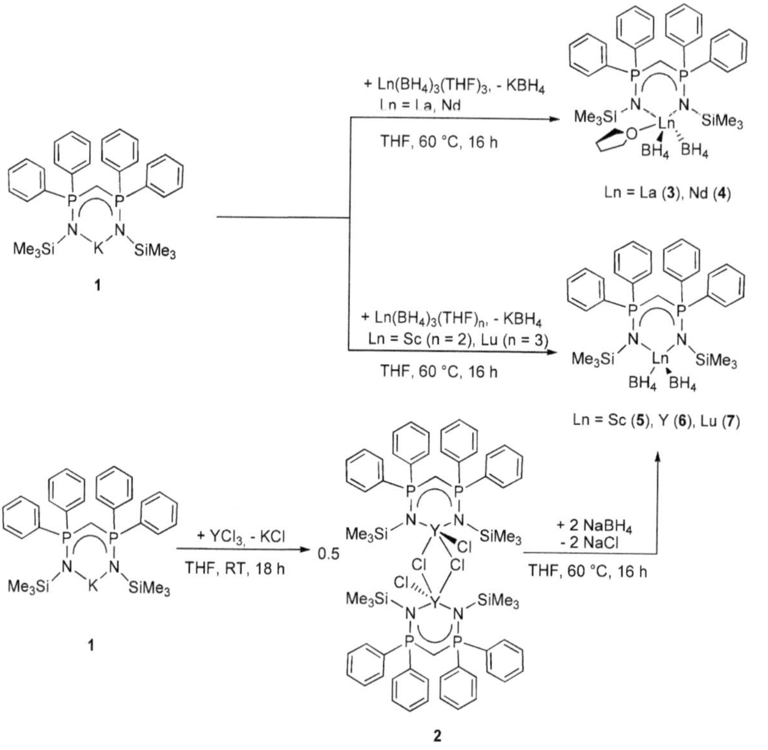

Scheme 18

The novel complexes **3-7** have been characterized by standard analytical / spectroscopic techniques and the solid state structures were established by single crystal X-ray diffraction. All complexes, except the paramagnetic neodymium compound **4**, have been characterized by NMR spectroscopy. The NMR spectra of **3**, **5**, **6** and **7** show the characteristic signals for the two different types of ligands. In the ^{1}H NMR spectra, one typical singlet for the SiMe$_3$ groups (δ = 0.22 (**3**), 1.09 (**5**), 0.19 (**6**) and 0.35 (**7**) ppm) and a triplet for the PCHP units (δ = 2.02 (**3**), 2.06 (**5**), 1.94 (**6**) and 1.83 (**7**) ppm) of the {CH(PPh$_2$NSiMe$_3$)$_2$}$^-$ ligands are observed. The ^{31}P{^{1}H} NMR spectra show one characteristic signal for the {CH(PPh$_2$NSiMe$_3$)$_2$}$^-$ ligand (δ = 15.6 (**3**), 19.9 (**5**) and 20.9 (**7**) ppm). For the yttrium compound **6** the signal is split into a doublet at 17.8 ppm as a result of the ^2J(P,Y) coupling.

The ^{1}H NMR spectra show the BH_4^- anions as broad signals (δ = 1.02-1.64 (**3**), 0.99-1.41 (**5**), 1.70-2.10 (**6**) and 1.45-2.08 (**7**) ppm). More characteristic are the ^{11}B NMR signals of these groups. The ^{11}B NMR spectra of compounds **3** and **6** show a quintet (δ = -22.4, $J_{B,H}$ = 83 Hz (**3**) and δ = -24.7, $J_{B,H}$ = 108 Hz (**6**)), while a broad signal for **7** was observed (δ = -25.8 ppm). In the EI-MS spectra, a molecular peak is observed as expected for **5-7**, while the mass spectra of **3** and **4** show the loss of a THF molecule and a borohydride anion. The IR spectra of **3-7** show two characteristic peaks for each complex at 2210 cm^{-1} and 2424 cm^{-1} (**3**), 2204 cm^{-1} and 2418 cm^{-1} (**4**), 2291 cm^{-1} and 2495 cm^{-1} (**5**), 2216 cm^{-1} and 2486 cm^{-1} (**6**) and 2225 cm^{-1} and 2420 cm^{-1} (**7**), which can be assigned to the coordinated borohydrides.[178]

Colorless needles of **3** suitable for single crystal X-ray diffraction analysis were obtained by crystallization from hot THF. Compound **3** crystallizes in the monoclinic space group $P2_1/n$ with four molecules of **3** and eight molecules of THF in the unit cell. As shown in Figure 1, {CH(PPh$_2$NSiMe$_3$)$_2$}$^-$ as an NCN tridentate donor ligand is coordinated by the two nitrogen atoms and the central carbon atom (C1) to the metal center (La-N1 2.552(3) Å, La-N2 2.556(3) Å and La-C1 2.789(3) Å). As expected, the La-C bond distance is longer than normally observed in organo lanthanum compounds.[35] By chelation of the {CH(PPh$_2$NSiMe$_3$)$_2$}$^-$ ligand, a six membered metallacycle (N1-P1-C1-P2-N2-La) is formed in a typical twist boat conformation, in which the central carbon atom and the lanthanum atom are displaced from the N$_2$P$_2$ least-square-plane. The BH_4^- groups are coordinated by the freely refined hydrogen atoms in a typical η^3-coordination mode.[107, 110, 114, 115, 178] The coordination polyhedron of **3** is formed by one {CH(PPh$_2$NSiMe$_3$)$_2$}$^-$ ligand, two BH_4^- anions and one THF molecule. If the BH_4^- group is considered as monodentate, the structure of a six-fold coordination sphere of the ligands around the metal atom is revealed in a distorted octahedral coordination polyhedron (e.g. B1-La-B2 100.8(2)°). The tetragonal base area of the distorted octahedron is formed by the THF molecule, the two nitrogen atoms of the {CH(PPh$_2$NSiMe$_3$)$_2$}$^-$ ligand and one BH_4^- anion, while the carbon atom of the ligand and the second BH_4^- anion form the apexes (B1-La-C1 152.1(2)°).

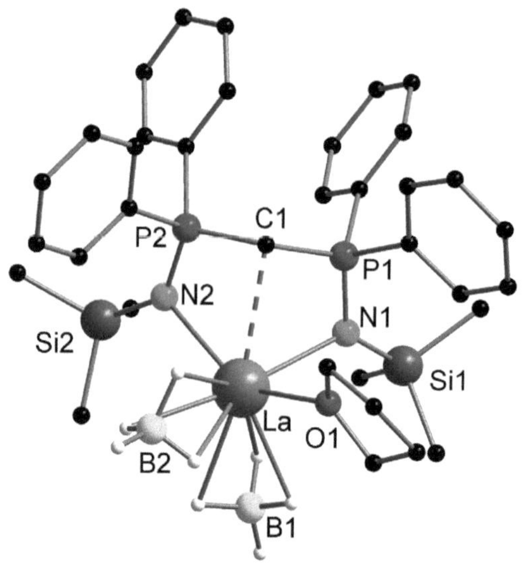

Figure 1 Solid state structure of **3**, omitting hydrogen atoms, except for the freely refined B-H atoms. Selected bond lengths [Å] or angles [°]: La-N1 2.552(3), La-N2 2.556(3), La-O1 2.540(3), La-B1 2.706(5), La-B2 2.727(6), La-C1 2.789(3), N1-P1 1.598(3), N1-Si1 1.734(3), N2-P2 1.599(3), N2-Si2 1.728(3), C1-P2 1.745(3), C1-P1 1.748(3); N1-La-N2 89.36(10), N1-La-O1 80.59(10), N2-La-O1 142.82(9), N1-La-B1 93.3(2), N2-La-B1 111.38(14), N1-La-B2 160.19(15), N2-La-B2 98.3(2), N1-La-C1 61.34(10), N2-La-C1 60.51(9), O1-La-B1 104.90(14), O1-La-B2 82.3(2), B1-La-B2 100.8(2), O1-La-C1 83.55(10), B1-La-C1 152.1(2), B2-La-C1 106.8(2), P1-C1-P2 125.5(2).

Compound **4** was obtained as pale purple plates by crystallization from THF/*n*-pentane and investigated *via* single crystal X-ray diffraction analysis. Compound **4** crystallizes in the triclinic space group *P*-1 with two molecules of **4** and two molecules of THF in the unit cell. The solid state structure of **4** is illustrated in Figure 2. As observed for **3**, compound **4** exhibits a distorted octahedral coordination polyhedron (e.g. B1-Nd-B2 100.15(11)° and B1-Nd-C1 152.47(9)°). Similarly to **3**, the {CH(PPh$_2$NSiMe$_3$)$_2$}$^-$ ligand in **4** is coordinated by the two nitrogen atoms and the central carbon atom (C1) to the metal center (Nd-N1 2.499(2) Å, Nd-N2 2.477(2) Å and Nd-C1 2.711(2) Å). In contrast to the dimeric monoborohydride complex [(C$_8$H$_8$)Nd(BH$_4$)(THF)]$_2$, in which the BH$_4^-$ groups act as bridging ligands, **4** forms a monomer in solid state.[111] The hydrogen atoms of the BH$_4^-$ groups of **4** show a η^3-coordination. This is consistent with the observed η^3-coordination mode of the BH$_4^-$ anions in the bisborohydride compound [{C$_5$(*i*Pr)$_5$}Nd(BH$_4$)$_2$(THF)] which is also monomeric in the solid state.[110]

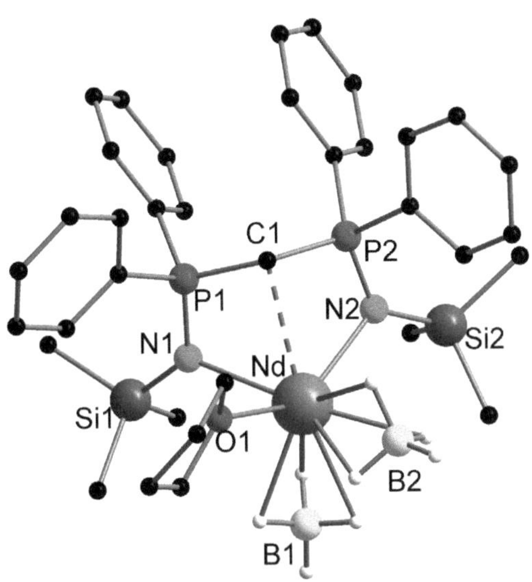

Figure 2 Solid state structure of **4**, omitting hydrogen atoms, except for the freely refined B-H atoms. Selected bond lengths [Å] or angles [°]: Nd-N1 2.499(2), Nd-N2 2.477(2), Nd-O1 2.4940(15), Nd-B1 2.642(3), Nd-B2 2.653(3), Nd-C1 2.711(2), N1-P1 1.599(2), N1-Si1 1.734(2), N2-P2 1.603(2), N2-Si2 1.736(2), C1-P2 1.750(2), C1-P1 1.754(2); N1-Nd-N2 91.40(6), N1-Nd-O1 80.29(5), N2-Nd-O1 141.07(5), N1-Nd-B1 93.65(9), N2-Nd-B1 106.64(8), N1-Nd-B2 159.69(8), N2-Nd-B2 98.75(8), N1-Nd-C1 62.93(6), N2-Nd-C1 62.65(5), O1-Nd-B1 111.78(8), O1-Nd-B2 80.78(7), B1-Nd-B2 100.15(11), O1-Nd-C1 79.99(5), B1-Nd-C1 152.47(9), B2-Nd-C1 106.42(8), P1-C1-P2 124.53(11).

Compound **5** was obtained as colorless plates by crystallization from hot toluene. The solid state structure of **5** was investigated by single crystal X-ray diffraction. Compound **5** crystallizes enantiomerically pure in the chiral monoclinic space group $P2_1$ with two molecules of **5** and two molecules of toluene in the unit cell. As a result of the smaller ionic radius, there is no solvent molecule coordinated to the metal center. Therefore, the coordination polyhedron of **5** is formed by the $\{CH(PPh_2NSiMe_3)_2\}^-$ ligand and the two BH_4^- anions (Figure 3), in this way building a five-fold coordination sphere around the metal atom which results in a distorted tetragonal pyramidal geometry (e.g. B1-Sc-B2 102.9(2)°). Similarly to **3** and **4**, a six membered metallacycle (N1-P1-C1-P2-N2-Sc) is formed by coordination of the $\{CH(PPh_2NSiMe_3)_2\}^-$ ligand, exhibiting a twist boat conformation. The bonds between the nitrogen atoms and the metal center (Sc-N1 2.134(3) Å and Sc-N2 2.160(3) Å) and the long interaction between the central carbon atom (C1) and the scandium atom (Sc-C1 2.471(3) Å) are shorter than the analogues bonds in the structures of **3** and **4**. This phenomenon is caused by the smaller ionic radius of the metal center in **5**, and as shown earlier for complexes which are coordinated by the $\{CH(PPh_2NSiMe_3)_2\}^-$ ligand, can also be influenced by the coordination sphere as well as by crystal packing effects.[43, 149] In contrast to all of the other complexes presented here, only one of the two BH_4^- anions is η^3-coordinated and the other one coordinates in an η^2-mode. The coordination mode is confirmed by the different Sc-B distances of the η^3-coordinated borohydride (Sc-B1 2.310(5) Å) and the η^2-coordinated borohydride (Sc-B2 2.526(5) Å). The η^2-coordination mode of the BH_4^- anion has been reported earlier in the scandium monoborohydride complex $[\{C_5H_3(SiMe_3)_2\}_2Sc(BH_4)]$ and the Sc-B distance (Sc-B 2.52(2) Å) is comparable to that observed for **5**.[179]

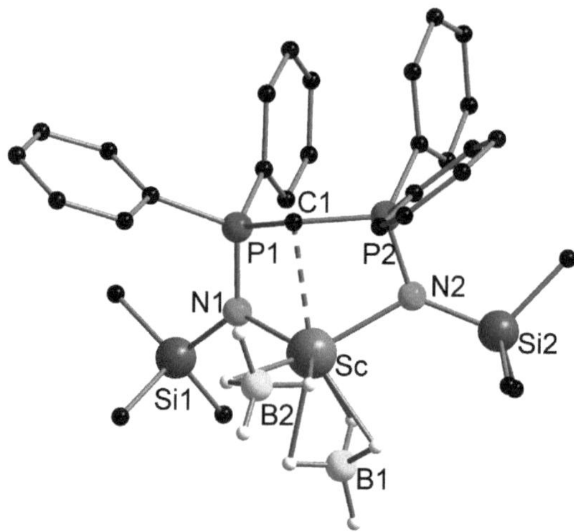

Figure 3 Solid state structure of **5**, measured at 200 K, omitting hydrogen atoms, except for the freely refined B-H atoms. Selected bond lengths [Å] or angles [°]: Sc-N1 2.134(3), Sc-N2 2.160(3), Sc-C1 2.471(3), Sc-B1 2.310(5), Sc-B2 2.526(5), N1-P1 1.609(3), N1-Si1 1.757(3), N2-P2 1.615(3), N2-Si2 1.755(3), C1-P2 1.737(4), C1-P1 1.748(4); N1-Sc-N2 111.97(12), N1-Sc-B1 103.3(2), N2-Sc-B1 102.3(2), N1-Sc-B2 114.1(2), N2-Sc-B2 119.6(2), N1-Sc-C1 70.66(12), N2-Sc-C1 69.58(12), B1-Sc-B2 102.9(2), B1-Sc-C1 165.8(2), B2-Sc-C1 91.3(2), P1-C1-P2 126.3(2).

In order to confirm the positions of the hydrogen atoms of the BH_4^- anions, a low temperature (9 K) single crystal X-ray diffraction of **5** was performed by G. Eickerling. As a result of reducing the thermal motion at low temperatures, the structural disorder and the smearing of the electron density are minimized. The positions of the hydrogen atoms close to the scandium center can be determined more exactly by subsequent structural refinements. Interestingly, by single crystal X-ray diffraction at low temperature a conformational polymorph of **5** was identified. Racemic crystals of compound **5** exhibit the centrosymmetric monoclinic space group $P2_1/c$ with four molecules of **5** and four molecules of toluene in the unit cell. Both BH_4^- groups are now η^3-coordinated (Sc-B1 2.350(3) Å and Sc-B2 2.355(2) Å), as shown in Figure 4. In the presented structure of **5** at 9 K, all hydrogen atoms are freely refined. Similarly to the structure of **5** at 200 K, the coordination polyhedron of **5** at 9 K shows a distorted tetragonal pyramidal geometry (e.g. B1-Sc-B2 108.65(9)°). The nitrogen atoms of the $\{CH(PPh_2NSiMe_3)_2\}^-$ ligand are coordinated symmetrically to the metal center with Sc-N bond distances (Sc-N1 2.152(2) Å and Sc-N2 2.148(2) Å) similar to those in **5** at 200 K (Sc-N1 2.134(3) Å and Sc-N2 2.160(3) Å). The interaction between the central carbon atom (C1) and the scandium atom (Sc-C1 2.543(3) Å) is slightly longer than the one obtained for **5** at 200 K (Sc-C1 2.471(3) Å). This could be a result of crystal packing effects by changing the space group. Furthermore, by changing the coordination mode of the BH_4^- anion from η^2 (at 200 K) to η^3 (at 9 K), the Sc-B bond distance is reduced (Sc-B2 2.526(5) Å (**5** at 200 K) and Sc-B2 2.355(2) (**5** at 9 K)).

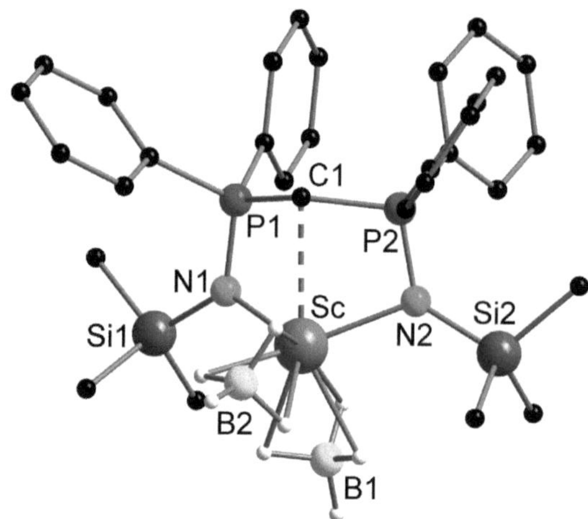

Figure 4 Solid state structure of **5**, measured at 9 K, omitting hydrogen atoms, except for B-H atoms (all hydrogen atoms are freely refined). Selected bond lengths [Å] or angles [°]: Sc-N1 2.152(2), Sc-N2 2.148(2), Sc-C1 2.543(3), Sc-B1 2.350(3), Sc-B2 2.355(2), N1-P1 1.609(2), N1-Si1 1.749(2), N2-P2 1.607(2), N2-Si2 1.751(2), C1-P2 1.738(2), C1-P1 1.731(2); N1-Sc-N2 117.91(7), N1-Sc-B1 96.88(8), N2-Sc-B1 98.90(9), N1-Sc-B2 114.54(8), N2-Sc-B2 116.01(8), N1-Sc-C1 69.02(7), N2-Sc-C1 68.98(7), B1-Sc-B2 108.65(9), B1-Sc-C1 151.39(8), B2-Sc-C1 99.92(8), P1-C1-P2 130.97(14).

Crystallization of **6** from hot THF gave colorless plates, while colorless disk shaped crystals of **7** were obtained by crystallization from toluene / *n*-pentane. Compounds **6** and **7** are isostructural and crystallize in the monoclinic space group $P2_1/n$ with four molecules of each complex in the unit cell. Figure 5 shows the structure of **6** and selected bond lengths and angles are given for **6** and **7**. There is no solvent molecule in the coordination sphere of **6** and **7**. Similarly to **5**, a distorted tetragonal pyramidal coordination polyhedron is observed for **6** and **7**. As expected, by chelation of the $\{CH(PPh_2NSiMe_3)_2\}^-$ ligands, six membered metallacycles (N1-P1-C1-P2-N2-Ln) are formed in the typical twist boat conformation.[152] The $\{CH(PPh_2NSiMe_3)_2\}^-$ ligands of **6** and **7** are coordinated by the two nitrogen atoms (Y-N1 2.297(2) Å and Y-N2 2.328(2) Å (**6**); Lu-N1 2.256(3) Å, Lu-N2 2.290(3) Å (**7**)) and by the central carbon atom (Y-C1 2.651(3) Å (**6**), Lu-C1 2.620(3) Å (**7**)). The hydrogen atoms of the BH_4^- anions are freely refined and show a similar η^3-coordination to **3** and **4**.

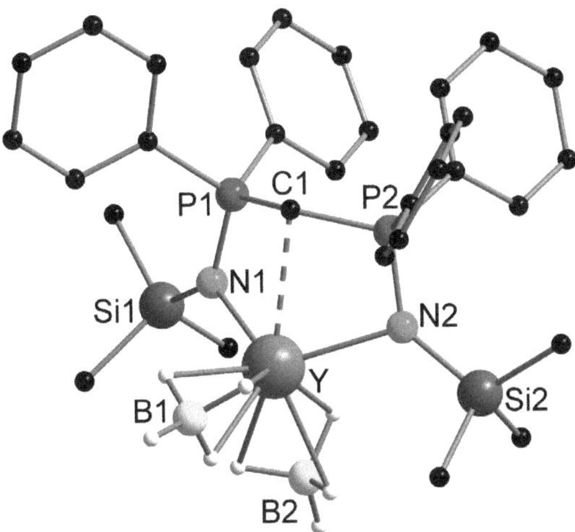

Figure 5 Solid state structure of **6**, omitting hydrogen atoms, except for the freely refined B-H atoms. Selected bond lengths [Å] or angles [°] are given for **6** and isostructural complex **7**.

6: Y-N1 2.297(2), Y-N2 2.328(2), Y-C1 2.651(3), Y-B1 2.496(4), Y-B2 2.500(4), N1-P1 1.611(2), N1-Si1 1.746(2), N2-P2 1.605(2), N2-Si2 1.742(2), C1-P2 1.736(3), C1-P1 1.746(3); N1-Y-N2 113.79(8), N1-Y-B1 116.61(13), N2-Y-B1 116.71(12), N1-Y-B2 96.50(12), N2-Y-B2 98.22(12), N1-Y-C1 65.98(8), N2-Y-C1 65.41(8), B1-Y-B2 111.2(2), B1-Y-C1 104.13(15), B2-Y-C1 144.66(13), P1-C1-P2 133.42(2).

7: Lu-N1 2.256(3), Lu-N2 2.290(3), Lu-B1 2.436(5), Lu-B2 2.449(5), Lu-C1 2.620(3), N1-P1 1.607(3), N1-Si1 1.748(3), N2-P2 1.601(3), N2-Si2 1.745(3), C1-P2 1.743(3), C1-P1 1.748(3); N1-Lu-N2 115.24(10), N1-Lu-B1 115.75(14), N2-Lu-B1 116.76(14), N1-Lu-B2 96.80(14), N2-Lu-B2 97.71(14), N1-Lu-C1 66.95(10), N2-Lu-C1 66.42(9), B1-Lu-B2 110.7(2), B1-Lu-C1 103.3(2), B2-Lu-C1 146.0(2), P1-C1-P2 132.3(2).

Catalytic applications

In collaboration with S. M. Guillaume *et al.*, the novel bis(phosphinimino)methanide rare earth metal borohydrides **3**, **6** and **7** were investigated for their catalytic activity in the ROP of CL.[177] All compounds are very active catalysts and allow the controlled ROP of CL in terms of molar mass and molar mass distribution with monomer-to-initiator ratios up to 300 (Table 2). The monomer conversions are nearly quantitative in all reactions and the polymer is recovered in 92 % yield or above with moderate to high molar mass. The reactions proceed easily in THF at room temperature and slightly less controlled in toluene (entries 4, 10, 13). The polymer with the highest molar mass is obtained from the reaction of the yttrium complex in toluene (entry 10, M_n = 22.4 $\times 10^3$ g/mol); but the molar mass distribution value is high, indicating an unselective reaction. The molar mass distribution of the polymers obtained at room temperature, although within a reasonable range, remain larger than those (typically 1.2-1.3) observed for the ROP of CL with [Ln(BH$_4$)$_3$(THF)$_3$] (Ln = La, Nd, Sm) and [(C$_5$Me$_5$)$_2$Sm(BH$_4$)(THF)].[128, 130, 131, 133] This suggests the occurrence of typical side reactions, such as transfer reactions and transesterification reactions,[180, 181] and can also be assigned to a propagation which is faster than the initiation. The control of the reactions was significantly improved by decreasing the reaction temperature (Table 2, entries 3, 9, 12). At 0 °C, the molar mass distribution values reached the narrowest values (1.06-1.11) ever obtained for the ROP of CL initiated from a rare earth metal borohydride species.[119, 128-131, 133-135] This indicates that at 0 °C the propagation is slowed down relative to the initiation, and thus the side reactions are limited. Complexes **3**, **6** and **7** showed similar activity with a monomer-to-initiator ratio of 300 or 250 at 20 °C (entries 2, 8, 11) and at 0 °C (entries 3, 9, 12). There is no dependence of the reactivity on the ionic radii of the metals observed. As illustrated by the TOF values reported in Table 2, all catalysts are highly active to form α,ω-dihydroxytelechelic polycaprolactones obtained by precipitation in methanol and analyzed *via* NMR spectroscopy.[177]

Table 2 Polymerization of CL initiated by **3**, **6** and **7**.[a]

Entry	Cat	$[CL]_0/[Cat]_0$	Solvent	Reaction temperature (°C)	Reaction time[b] (min)	Conv.[c] (%)	$M_{n,theo}$[d] ($\times 10^3$ g/mol)	$M_{n,SEC}$[e] ($\times 10^3$ g/mol)	M_w/M_n[f]	TOF (mol_{CL} $mol_{Catalyst}^{-1} \cdot h^{-1}$)
1	**3**	50	THF	20	2	100	2.8	2.2	1.16	1 500
2	**3**	300	THF	20	1	98	16.8	13.4	1.38	17 640
3	**3**	300	THF	0	20	95	16.2	15.7	1.11	855
4	**3**	220	toluene	20	0.5	95	11.9	12.9	1.47	25 080
5	**6**	50	THF	20	2	100	2.8	2.6	1.18	1 500
6	**6**	100	THF	20	5	100	5.7	5.6	1.40	1 200
7	**6**	250	THF	20	100	100	14.3	15.1	1.45	150
8	**6**	300	THF	20	1	92	16.2	15.7	1.49	16 560
9	**6**	250	THF	0	30	98	14.0	14.3	1.09	490
10	**6**	250	toluene	20	300	99	14.1	22.4	1.84	50
11	**7**	300	THF	20	1	98	16.8	15.3	1.35	17 640
12	**7**	300	THF	0	20	96	16.4	15.8	1.06	864
13	**7**	300	toluene	20	0.5	95	16.2	14.2	1.48	34 200

[a]Results are representative of at least duplicated experiments. [b]Reaction times were not necessarily optimized. [c]Monomer conversion determined by ^{1}H NMR. [d]Theoretical molar mass calculated from $[CL]_0/2[Cat]_0 \times$ monomer conversion $\times M_{CL}$, with M_{CL} =114 g.mol[-1]. [e]Experimental molar mass determined by SEC *versus* polystyrene standards and corrected by a factor of 0.56.[130] [f]Molar mass distribution calculated from SEC traces.

In addition to the experimental data of the ROP of CL collected by S. M. Guillaume *et al.*, L. Maron *et al.* performed a computational study of the reaction of **6** with CL in order to gain more insights into the reaction mechanism.[177] In order to facilitate the calculations, the {CH(PPh₂NSiMe₃)₂}⁻ ligand has been simplified by replacing the phenyl ring by a methyl group and the SiMe₃ unit by SiH₃. The calculated free energy profile for the reaction of **6** with CL is shown in Figure 6.

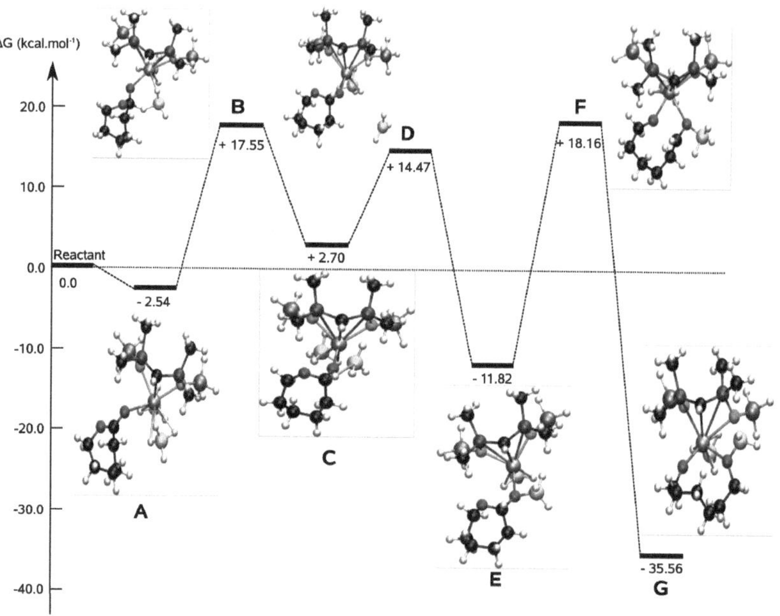

Figure 6 Calculated free-energy profile for the reaction of ε-caprolactone with **6**.

The calculated reaction mechanism, illustrated in Scheme 19, is very similar to that previously proposed for the monoborohydride complexes [(C₅H₅)₂Eu(BH₄)] and [(N₂NN')-Eu(BH₄)] (N₂NN'=(2-C₅H₄N)CH₂N(CH₂-CH₂NR)₂, R = SiMe₃ or 2,4,6-C₆H₂Me₃).[126] The initial step of the reaction of all three different types of complexes with CL is the coordination of the carbonyl oxygen atom of the CL molecule to the metal center to form **A**, followed by an interaction of the hydride with the carbonyl carbon atom (**B**) (Scheme 19). By using the previously presented complexes [(Lₓ)Ln(BH₄)] (Lₓ = (C₅H₅)₂, (N₂NN')) (Scheme 19, pathway b)), calculations showed that **B** is directly transformed into **E**, by hydride transfer from the rare earth metal complex to the carbonyl carbon atom of the CL. In the alkoxide-borate complex **E**, the BH_3 unit is coordinated to the exocyclic oxygen atom and additionally interacts with the metal center through two hydrogen atoms. In contrast, calculations of the present work involving the bis(phosphinimino)methanide ligand show that the formation of the borate ($-C(O)HBH_3$) **E** is achieved in two steps starting from **B** (Scheme 19, pathway a)). In the first step, the nucleophilic attack of the BH_4^- hydride at the carbonyl carbon atom of the CL leads to the release of BH_3 which remains coordinated to the C-H hydrogen atom to form **C** (Scheme 19, Figure 6). This unexpected formation of adduct **C** requires the cleavage of a strong electrostatic interaction between a borohydride ligand and the metal center. However, the formation of **C** is predicted to be only slightly endergonic (Figure 6). In contrast to the previously studied monohydrides, the bis(phosphinimino)-methanide compounds of the present work are bisborohydride complexes. Adduct **C** is stabilized by an interaction between the BH_3 and the other BH_4^- ligand in a donor-acceptor way (interaction between a hydride of BH_4^- and the empty p orbital of BH_3). The second BH_4^- ligand remains bonded to the metal center in all steps of the reaction. In addition, the loss of the electrostatic interaction between a borohydride ligand and the metal center in **C** is compensated by the high donor capability of the bis(phosphinimino)methanide ligand. The formation of the borate **E** occurs in the second step (Scheme 19, pathway a), in which the BH_3 is trapped by the oxygen of the ketone.

The following steps of the mechanism are similar to that previously reported for $[(L_x)Ln(BH_4)]$ ($Lx = (C_5H_5)_2$, (N_2NN')). The alkoxyborane group in **G** is formed by the reduction of the carbonyl carbon atom of the CL by the BH_3. The propagation step involves the coordination-insertion of further CL molecules through the Ln-O bond to give the polymer **H**. At last, the hydrolysis of the Ln-O and of the $-CH_2O-BH_2$ bonds of **H** leads to the formation of α,ω-dihydroxytelechelic polycaprolactones. The reaction is predicted to be both kinetically and thermodynamically accessible.

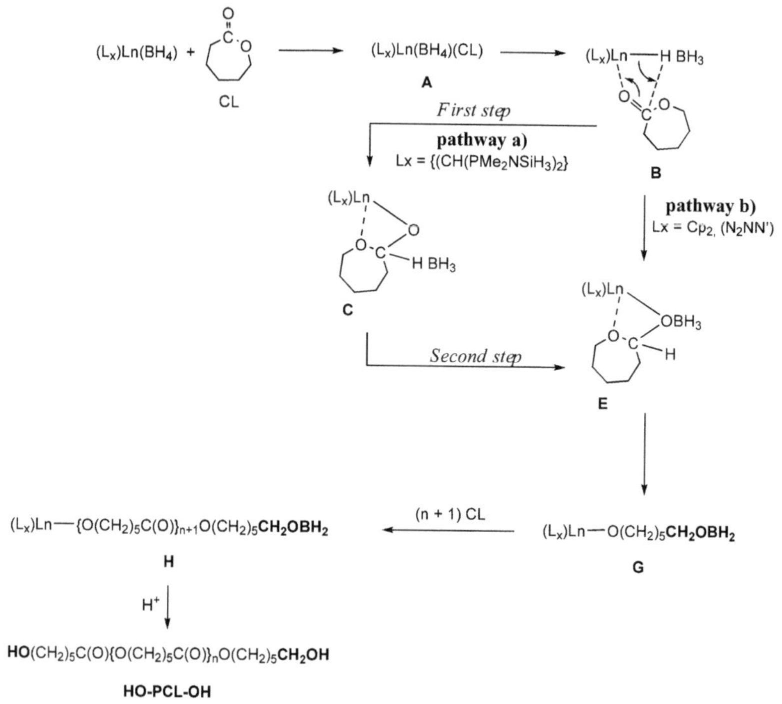

$(L_x)Ln(BH_4)$ + CL $\longrightarrow$ $(L_x)Ln(BH_4)(CL)$ **A** $\longrightarrow$ $(L_x)Ln$—H BH_3 **B**

First step

pathway a)
Lx = {(CH(PMe_2NSiH_3)_2}

C

pathway b)
Lx = Cp_2, (N_2NN')

E

Second step

$(n + 1)$ CL

$(L_x)Ln$—$O(CH_2)_5CH_2OBH_2$ **G**

$(L_x)Ln$—$\{O(CH_2)_5C(O)\}_{n+1}O(CH_2)_5CH_2OBH_2$ **H**

H^+

$HO(CH_2)_5C(O)\{O(CH_2)_5C(O)\}_nO(CH_2)_5CH_2OH$

HO-PCL-OH

Scheme 19 Proposed general mechanism for the polymerization of ε-caprolactone initiated by $[(L_x)Ln(BH_4)_n]$ with $L_x = (C_5H_5)_2$, (N_2NN'), n = 1 [126] and $L_x = \{(CH(PMe_2NSiH_3)_2\}$, n = 2.

In summary, bis(phosphinimino)methanide bisborohydrides of various rare earth metals were synthesized. They exhibit interesting coordination modes in the solid state depending on the ionic radii of the metal center and on the temperature. Complexes **3**, **6** and **7** successfully allowed the ring-opening polymerization of CL with nearly quantitative monomer conversion. In fact, at 0 °C, the molar mass distribution values reached the narrowest values ever obtained for the ROP of CL initiated by a rare earth metal borohydride species. In addition, computational studies showed a new mechanistic pathway which confirmed the good donor capability of the bis(phosphinimino)methanide ligand.

3.2   Complexes of the 2,5-bis{*N*-(2,6-diisopropylphenyl)iminomethyl}pyrrolyl ligand

3.2.1 2,5-Bis{*N*-(2,6-diisopropylphenyl)iminomethyl}pyrrolyl alkali metal complexes

As described in Section 1.5, the lithium and the potassium complexes [(DIP$_2$pyr)M] (M = Li (**9**), K (**10**)) were synthesized as ligand transfer reagents for salt metathesis reactions.[160, 163] These compounds have been characterized by the standard analytical / spectroscopic techniques, but the solid state structures have not been investigated. In the presented work, the lithium derivative and the hitherto unknown sodium complex [(DIP$_2$pyr)M]$_2$ (M = Li (**9**), Na (**11**)) were prepared and crystallization afforded crystals suitable for single crystal X-ray diffraction analysis. Interestingly, both compounds form dimeric complexes in the solid state, in which two metal ions are coordinated by two (DIP$_2$pyr)$^-$ ligands.[182]

As shown in Scheme 20, the reaction of 2,5-bis{*N*-(2,6-diisopropylphenyl)iminomethyl}-pyrrole (**8**) with *n*BuLi in toluene afforded the desired complex [(DIP$_2$pyr)Li]$_2$ (**9**). The novel sodium compound **11** was obtained by the reaction of **8** with NaH in hot THF (Scheme 20).

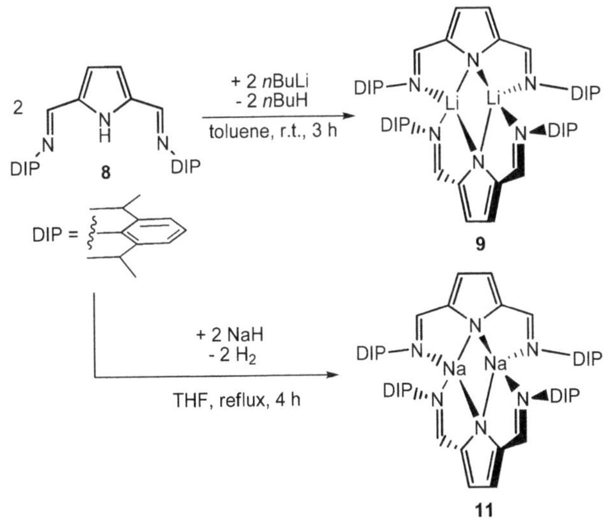

Scheme 20

Both compounds **9** and **11** were crystallized from hot toluene to give yellow crystals and the solid state structures were analyzed by single crystal X-ray diffraction. In addition, the known complex **9** and the novel complex **11** were characterized by standard analytical / spectroscopic techniques. In the ^{1}H NMR spectra one doublet (δ = 1.18 (**9**), 1.22 (**11**) ppm) and a clear septet (δ = 3.22 (**9**), 3.26 (**11**) ppm) are observed for the isopropyl groups of the (DIP$_2$pyr)$^-$ ligand. This suggests that, in contrast to other metal complexes coordinated by the (DIP$_2$pyr)$^-$ ligand,[163] the 2,6-diisopropylaniline moieties of the ligand in **9** and **11** can freely rotate in solution. For the imino N=CH units one signal (δ = 8.18 (**9**), 7.98 (**11**) ppm) is observed, for complex **9** only when C$_6$D$_6$ is used as solvent. This additionally indicates the symmetric coordination of the ligand. In contrast, for compound **9** a multiplet is observed in THF-d$_8$ (δ = 7.95 ppm), which implies that an interaction with the solvent takes place, leading to an asymmetric coordination.

Compound **9** crystallizes in the monoclinic space group *C2/c* with sixteen dimeric molecules in the unit cell (V = 22359(8) Å^3) and with two independent dimeric molecules in the asymmetric unit. As shown in Figure 7, two lithium atoms are coordinated by two (DIP$_2$pyr)$^-$ ligands which show η^3-coordination *via* all nitrogen atoms. The coordination polyhedron of each single lithium atom is formed by four nitrogen atoms, two of each ligand (Li1: N2-N3-N6-N5; Li2: N1-N2-N5-N4), resulting in a distorted tetrahedral coordination geometry (e.g. N1-Li2-N2 86.007(12)°, N2-Li1-N3 86.663(12)°). The pyrrolyl units of both ligands are bridging the two lithium atoms and the coordination tetrahedrons of both lithium atoms are edge-linked, forming a diamond-shaped Li$_2$N$_2$ core. The distance between the two lithium atoms is 2.5653(5) Å. The whole structure can be described by four five membered rings (Li-N-C-C-N), based on the Li$_2$N$_2$ core. This structural motif is very rare in lithium chemistry.[183-185]

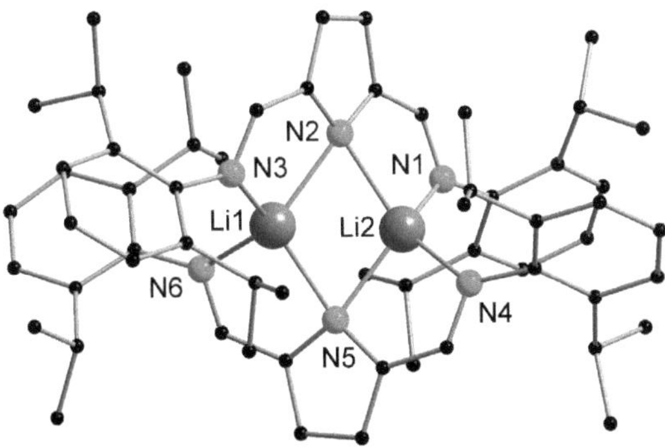

Figure 7 Solid state structure of **9** showing the atom labeling scheme, omitting hydrogen atoms. Only one of two independent molecules is shown. Selected bond lengths [Å] or angles [°]: Li1-N2 2.1934(4), Li1-N3 2.0265(3), Li1-N5 2.2019(5), Li1-N6 2.0244(4), Li2-N1 2.0194(4), Li2-N2 2.1994(5), Li2-N4 2.0464(4), Li2-N5 2.1612(4), Li3-N7 2.0501(4), Li3-N8 2.1919(4), Li3-N10 2.0611(4), Li3-N11 2.2103(4), Li4-N8 2.1738(4), Li4-N9 2.0502(4), Li4-N11 2.1650(4), Li4-N12 2.0217(4), Li1-Li2 2.5653(5), Li3-Li4 2.5587(5); N2-Li1-N3 86.663(12), N2-Li1-N6 130.540(14), N2-Li1-N5 107.574(15), N3-Li1-N5 138.780(15), N3-Li1-N6 114.43(2), N5-Li1-N6 85.814(13), N1-Li2-N2 86.007(12), N1-Li2-N4 116.83(2), N1-Li2-N5 131.918(15), N2-Li2-N4 133.36(2),N4-Li2-N5 86.209(12), N5-Li2-N2 108.823(15), N7-Li3-N10 117.73(2), N7-Li3-N8 86.221(12), N7-Li3-N11 133.565(15), N8-Li3-N10 133.301(15), N8-Li3-N11 107.184(15), N10-Li3-N11 85.318(12), N8-Li4-N9 86.085(12), N8-Li4-N11 109.486(15), N8-Li4-N12 131.839(15), N9-Li4-N11 131.731(15), N9-Li4-N12 117.56(2), N11-Li4-N12 86.125(13).

Compound **11** crystallizes in the triclinic space group *P*-1 with two molecules of **11** and three molecules of toluene in the unit cell. Since lithium and sodium cations have a different ionic radius, the structures of compound **9** and **11** are related but not isostructural. Similar to **9**, complex **11** is dimeric in the solid state and two sodium atoms are coordinated by two (DIP₂pyr)⁻ ligands, each in a η^3-coordination mode (Figure 8). The (DIP₂pyr)⁻ ligands are arranged almost perpendicular to each other. Each sodium atom is coordinated by four nitrogen atoms, forming a distorted tetrahedron. This distortion is significantly larger than in compound **9**, which derives from the larger ionic radius of the sodium ion compared to the lithium ion (e.g. N1-Na2-N2 70.69(11)°, N2-Na1-N3 77.65(11)°). The pyrrolyl units of both ligands are bridging between the sodium atoms (Na1-Na2 3.045(2) Å), forming a diamond-shaped Na_2N_2 core. As observed for compound **9**, the whole structure can be best described as four five membered rings (Na-N-C-C-N), based on the Na_2N_2 core. This kind of structural motif is hitherto unknown for sodium compounds.

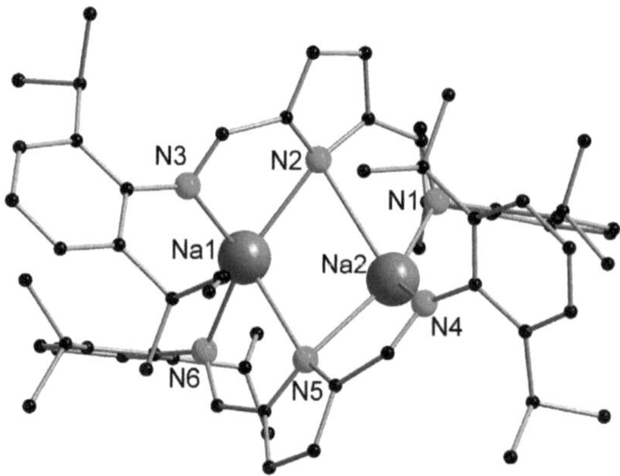

Figure 8 Solid state structure of **11** showing the atom labeling scheme, omitting hydrogen atoms. Selected bond lengths [Å] or angles [°]: Na1-N2 2.451(3), Na1-N3 2.369(3), Na1-N5 2.513(3), Na1-N6 2.382(3), Na2-N1 2.413(3), Na2-N2 2.800(4), Na2-N4 2.404(3), Na2-N5 2.492(3), Na1-Na2 3.045(2); N2-Na1-N3 77.65(11), N2-Na1-N5 107.71(11), N2-Na1-N6 148.89(12), N3-Na1-N5 119.51(12), N3-Na1-N6 129.42(12), N5-Na1-N6 74.44(11), N1-Na2-N2 70.69(11), N1-Na2-N4 132.28(12), N1-Na2-N5 150.05(11), N2-Na2-N4 107.04(11), N2-Na2-N5 98.30(11), N4-Na2-N5 77.21(10).

3.2.2 2,5-Bis{*N*-(2,6-diisopropylphenyl)iminomethyl}pyrrolyl rare earth metal chlorides and borohydrides

As described in chapter 1.3, rare earth metal borohydrides found widespread applications as efficient catalysts for the polymerization and copolymerization of styrene,[120-122] isoprene[112, 114, 120, 123, 124] and ethylene.[114, 125] In addition, neodymium-based Ziegler-Natta systems play a major role in the industrial polymerization of 1,3-butadiene to poly-*cis*-1,4-butadiene.[186-189] Poly-*cis*-1,4-butadiene is an important component of standard rubber material used for the production of tire treads.[190, 191] Ziegler-Natta systems, containing neodymium halides or carboxylates together with aluminium derivatives, have been well established, showing high activities and *cis*-selectivities in the polymerization of 1,3-butadiene.[190] Furthermore, (η^3-allyl)-neodymium complexes such as [(η^3-C$_3$H$_5$)$_3$Nd(dioxane)] and [(η^3-C$_3$H$_5$)$_2$Nd(μ-Cl)(THF)]$_2$ as well as [(η^3-C$_3$H$_5$)NdCl$_2$(THF)$_2$] have shown to be extremely active and highly selective catalysts.[192-194] Cui *et al.* synthesized aryldiimine NCN-pincer ligated rare earth metal dichlorides [{2,6-(2,6-C$_6$H$_3$R$_2$N=CH)$_2$C$_6$H$_3$}LnCl$_2$(THF)$_2$]. By using aluminum tris(alkyl) derivatives and [Ph$_3$C][B(C$_6$F$_5$)$_4$] as cocatalysts, the new compounds showed high activities and *cis*-1,4 selectivities in the polymerization of 1,3-butadiene.[195] As described in chapter 1.5, the 2,5-bis{*N*-(2,6-diisopropylphenyl)iminomethyl}pyrrolyl ligand was proved to be very suitable in stabilizing rare earth metal complexes.[158, 163] In the presented work, 2,5-bis{*N*-(2,6-diisopropylphenyl)iminomethyl}pyrrolyl rare earth metal chlorides and borohydrides were synthesized and two of the neodymium containing species were investigated for the polymerization of 1,3-butadiene.

In collaboration with N. Meyer from our group, who already synthesized the chlorides of the smaller rare earth metals [(DIP$_2$pyr)LnCl$_2$(THF)$_2$] (Ln = Y, Lu),[163, 164] the 2,5-bis{*N*-(2,6-diisopropylphenyl)iminomethyl}pyrrolyl dichloride complex of neodymium was prepared in order to explore the catalytic activity in the polymerization of 1,3-butadiene.[196] The neodymium complex **12** was obtained by the reaction of the potassium salt **10** with anhydrous NdCl$_3$ in THF at 60 °C (Scheme 21). [(DIP$_2$pyr)NdCl$_2$(THF)]$_2$ (**12**) is dimeric in the solid state, while the chloride complexes of the smaller rare earth metal ions are monomeric.

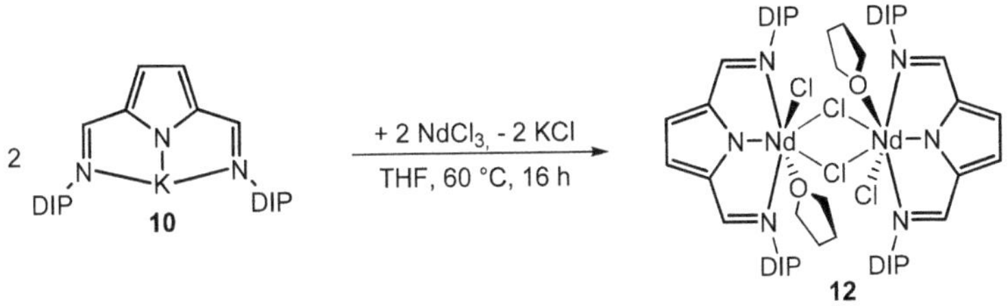

Scheme 21

The solid state structure of the paramagnetic compound **12** was established by single crystal X-ray diffraction. Compound **12** crystallizes in the triclinic space group *P*-1 with one molecule of **12** and two molecules of THF in the unit cell. As shown in Figure 9, **12** is dimeric and the metal centers are bridged almost symmetrically by two chlorine atoms (Nd-Cl2 2.807(2) Å, Nd-Cl2′ 2.825(2) Å). Similar structural motifs are well established in lanthanide chemistry.[35, 154, 197-199] In the center of the Nd-Cl2-Nd′-Cl2′ plane, a crystallographic inversion center is localized. As observed for the yttrium and lutetium analogues,[163] the (DIP$_2$pyr)⁻ ligand in compound **12** is symmetrically attached to the metal center, showing a η^3-coordination mode. The metal-imine nitrogen bond distances (Nd-N1 2.715(7) Å and Nd-N3 2.713(7) Å) are longer than the metal-pyrrolyl distance (Nd-N2 2.369(7) Å). The seven-fold coordination sphere around each metal center is formed by the (DIP₂pyr)⁻ ligand, two chlorine atoms and one THF molecule, resulting in a distorted pentagonal bipyramidal coordination polyhedron. The THF molecule and one chlorine atom exhibit an almost linear setup (O1-Nd-Cl1 173.62(15)°). The sum of the corresponding five valence angles (358.64°) of the LnN₃(μ-Cl)₂ fragment shows an almost planar arrangement.

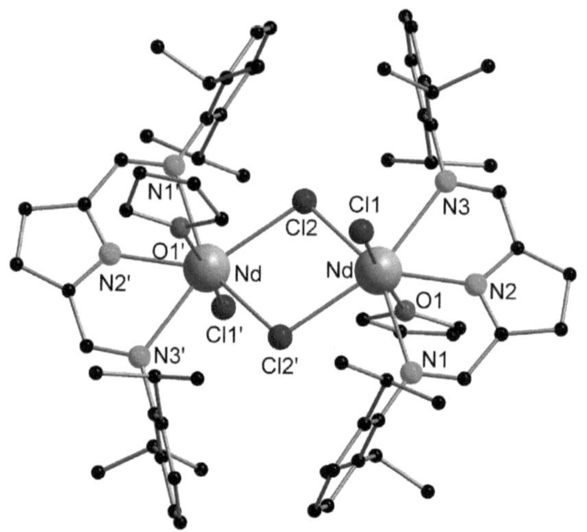

Figure 9 Solid state structure of **12** showing the atom labeling scheme, omitting hydrogen atoms. Selected bond lengths [Å] or angles [°]: Nd-N1 2.715(7), Nd-N2 2.369(7), Nd-N3 2.713(7), Nd-O1 2.491(6), Nd-Cl1 2.625(2), Nd-Cl2 2.807(2), Nd-Cl2′ 2.825(2); N1-Nd-N2 63.3(2), N1-Nd-N3 126.2(2), N1-Nd-O1 85.5(2), N1-Nd-Cl1 90.64(15), N1-Nd-Cl2 152.28(15), N1-Nd-Cl2′ 77.48(15), N2-Nd-N3 62.9(2), N2-Nd-O1 78.7(2), N2-Nd-Cl1 95.1(2), N2-Nd-Cl2 139.3(2), N2-Nd-Cl2′ 138.6(2), N3-Nd-O1 86.2(2), N3-Nd-Cl1 91.96(15), N3-Nd-Cl2 79.07(15), N3-Nd-Cl2′ 154.28(15), O1-Nd-Cl1 173.62(15), O1-Nd-Cl2 85.10(15), O1-Nd-Cl2′ 85.9(2), Cl1-Nd-Cl2 100.59(7), Cl1-Nd-Cl2′ 98.25(7), Cl2-Nd-Cl2′ 75.89(6), Nd-Cl2-Nd′ 104.11(6).

Furthermore, in collaboration with N.Meyer rare earth metal borohydride complexes of the 2,5-bis{*N*-(2,6-diisopropylphenyl)iminomethyl}pyrrolyl ligand were synthesized.[164, 196, 200] As shown in Scheme 22, the borohydride complexes **13-16** were prepared by the reaction of [(DIP₂pyr)K] (**10**) with the corresponding trisborohydrides [Ln(BH₄)₃(THF)ₙ] (Ln = Sc, n = 2; Ln = La, Nd, Lu, n = 3) in THF at elevated temperature. As described in chapter 3.1, rare earth metal trisborohydrides are easily synthesized by the reaction of anhydrous LnCl₃ with NaBH₄[108] and consequently, the desired products **13-16** were obtained in a convenient two step synthesis. The reaction of **10** with the trisborohydrides of the larger rare earth metals [Ln(BH₄)₃(THF)₃] (Ln = La, Nd) afforded the products [(DIP₂pyr)Ln(BH₄)₂(THF)₂]

(Ln = La (**13**), Nd (**14**)). As expected, the trisborohydrides react as pseudo halide compounds forming KBH_4 as byproduct (Scheme 22). In contrast, by using the trisborohydrides of the smaller rare earth metals $[Ln(BH_4)_3(THF)_n]$ (Ln = Sc (n = 2); Ln = Lu (n = 3)), a redox reaction of the BH_4-group with one of the Schiff-base functions of the ligand is observed. As a result, a dinegatively charged ligand with a new amido function is formed, which coordinates *via* the amido and the pyrrolyl unit to the metal center. In addition, the by-product of the reaction, a BH_3 molecule, is trapped in a unique reaction mode in the coordination sphere of the metal complex. The nitrogen atom of the remaining Schiff-base function of the ligand is bound to the BH_3 molecule, forming a =N-BH_3 group. The =N-BH_3 unit coordinates *via* two hydrogen atoms to the metal center. The resulting products $[\{DIP_2pyr^*-BH_3\}Ln(BH_4)(THF)_2]$ (Ln = Sc (**15**), Lu (**16**)) are shown in Scheme 22.

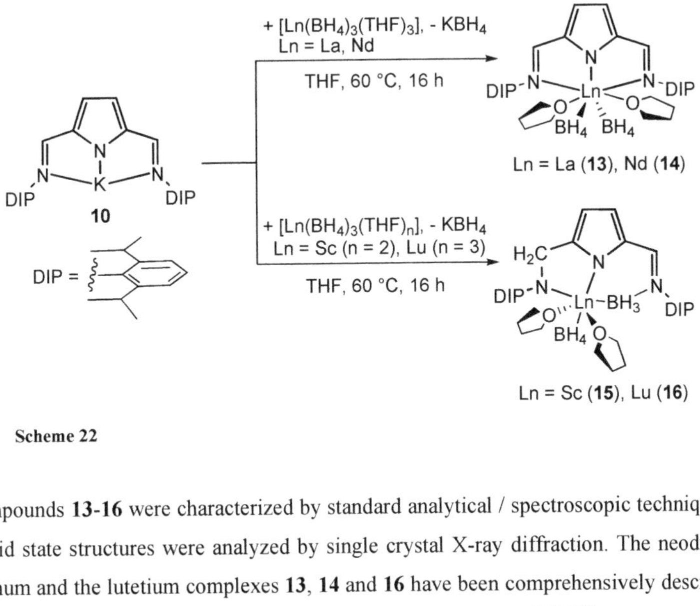

Scheme 22

Compounds **13-16** were characterized by standard analytical / spectroscopic techniques and the solid state structures were analyzed by single crystal X-ray diffraction. The neodymium, lanthanum and the lutetium complexes **13**, **14** and **16** have been comprehensively described in the PhD thesis of N. Meyer[164] and in the corresponding literature[196, 200] and will briefly be discussed here.

The ^1H and $^{13}C\{^1H\}$ NMR spectra of the diamagnetic compound **13** show the expected sets of signals of the symmetrically coordinated $(DIP_2pyr)^-$ ligand, and in the ^{11}B NMR spectrum a well resolved quintet at δ -21.3 ppm for the BH_4^- groups is observed.[164, 200] As a result of the reduction process, the symmetry of the ligands of compounds **15** and **16** is broken, which can be observed in the ^1H and $^{13}C\{^1H\}$ NMR spectra. In the ^1H NMR spectrum of complex **16** the four isopropyl CH_3 groups of the $(DIP_2pyr^*)^{2-}$ ligand give four doublets (δ = 1.12, 1.14, 1.15, 1.21 ppm), whereas for compound **15** a broad multiplet is observed (δ = 1.10-1.31 ppm). For both compounds two septets for the isopropyl CH groups (δ = 3.17, 3.72 (**15**); 3.05, 3.73 (**16**) ppm) and two doublets for the pyrrolyl ring (δ = 6.24, 7.08 (**15**); 6.21, 7.04 (**16**) ppm) are observed. The N-CH_2 moiety, formed by the reduction process, gives the expected singlet in the ^1H NMR spectrum for each complex (δ = 4.46 (**15**), 4.51 (**16**) ppm). The $^{13}C\{^1H\}$ NMR spectra of both compounds confirm these observations. The ^{11}B NMR spectra show well separated signals for the BH_4^- group (δ = -24.6 (**15**), -25.2 (**16**) ppm) and for the =N-BH_3 moiety (δ = -16.5 (**15**), -14.9 (**16**) ppm).

Compounds **13** and **14** are isostructural and crystallize in the orthorhombic space group *Pbca* with eight molecules in the unit cell. As an example, the solid state structure of the neodymium complex **14** is illustrated in Figure 10.[164, 200] The $(DIP_2pyr)^-$ ligand is symmetrically coordinated, and the BH_4^- anions show the expected η^3-coordination mode. In contrast to the neodymium complex **12** described above, which is μ-coordinated by two chlorine atoms and thus dimeric in solid state, the BH_4^- groups in compounds **13** and **14** are not acting as bridging ligands. Compounds **13** and **14** are monomeric in solid state and the coordination sphere of the metal center is satisfied by two additionally coordinated THF molecules.

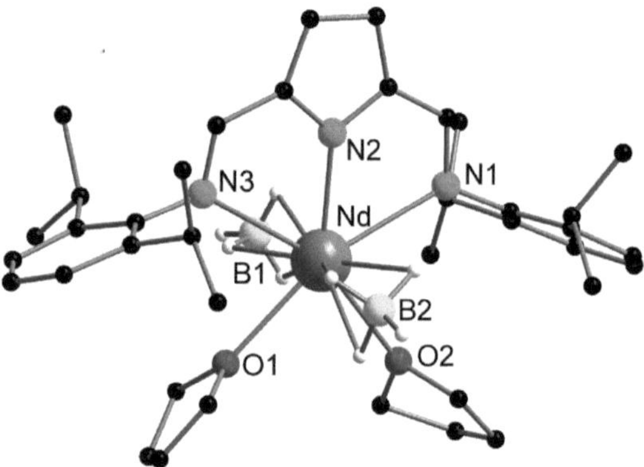

Figure 10 Solid state structure of **14** showing the atom labeling scheme, omitting hydrogen atoms, except for the freely refined B-H atoms.[164, 200]

Compounds **15** and **16** are also isostructural and crystallize in the monoclinic space group *Pn* with two molecules of **15** or **16** and four molecules of THF in the unit cell. The solid state structure of the scandium complex **15** is shown in Figure 11, while selected bond lengths or angles for **16** are given in the literature.[164, 200] As described above, the ligand of **15** and **16** exhibits a new amido function and the remaining Schiff-base function, which differ significantly in their C-N bond lengths (N1-C1 1.306(5) Å vs. N3-C6 1.473(4) Å (**15**) and N1-C1 1.317(5) Å vs. N3-C6 1.465(5) Å (**16**)), showing the single bond character of the amido $N-CH_2$ moiety. The BH_3 group is coordinated to the imino function, featuring a covalent B-N bond length of B2-N1 1.570(5) Å (**15**) and 1.562(6) Å (**16**). The $=N-BH_3$ moiety is coordinated in a η^2-fashion to the metal center, while the BH_4^- anion is η^3-coordinated (Sc-B1 2.365(5) Å, Sc-B2 2.845(5) Å (**15**) and Lu-B1 2.486(5) Å, Lu-B2 2.853(5) Å (**16**)). The coordination polyhedra of **15** and **16** are formed by the two nitrogen atoms of the pyrrolyl ligand (N2, N3), the BH_3 molecule, the BH_4^- anion and two THF molecules. By considering the B-H ligands as monodentate, the resulting six-fold coordination spheres of both compounds show a distorted octahedral geometry. The solid state structures of both compounds were determined at 200 K by single crystal X-ray diffraction. In addition, the solid state structure of complex **16** was determined again at low temperature (6 K) to minimize the smearing of the electron density due to thermal motion and to overcome the structural disorder observed above 200 K. The positions of the hydrogen atoms close to the lutetium center can be determined exactly by subsequent structural refinements. The η^2-coordination mode of the $=N-BH_3$ moiety as well as the η^3-coordination mode of the BH_4^- ligand were confirmed by this method.

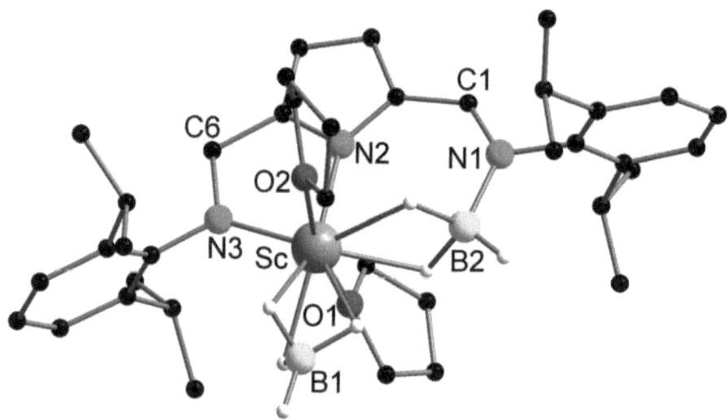

Figure 11 Solid state structure of **15** showing the atom labeling scheme, omitting hydrogen atoms, except for the freely refined B-H atoms. Selected bond lengths [Å] or angles [°]: Sc-N2 2.186(3), Sc-N3 2.041(3), Sc-O1 2.293(3), Sc-O2 2.285(3), Sc-B1 2.365(5), Sc-B2 2.845(5), N1-C1 1.306(5), N3-C6 1.473(4); N2-Sc-N3 79.53(11), N2-Sc-O1 87.23(11), N2-Sc-O2 84.79(11), O1-Sc-B1 92.9(2), O1-Sc-B2 84.52(15), O2-Sc-B1 94.2(2), N2-Sc-B1 175.0(2), N2-Sc-B2 75.02(12), N3-Sc-B1 105.5(2), N3-Sc-B2 154.54(12), B1-Sc-B2 100.0(2).

In the presented work, the successful synthesis of the scandium complex **15** being isostructural to **16** confirms that the reactivity of [(DIP$_2$pyr)K] (**10**) with [Ln(BH$_4$)$_3$(THF)$_n$] depends on the ionic radii of the center metals. A redox reaction of a BH$_4$-group with one of the Schiff-base functions of the ligand occurs only for the smaller rare earth metals. The reason for that unusual reactivity is unclear. The difference in reactivity should not be a result of the number of coordinated solvent molecules, since the trisborohydrides of lanthanum and neodymium as large metal centers and lutetium as a small metal center [Ln(BH$_4$)$_3$(THF)$_3$][107] exhibit the same number of coordinated solvent molecules (all compounds, except [Sc(BH$_4$)$_3$(THF)$_2$][201, 202]). Probably, an interaction between the coordinated BH$_3$ molecule and the BH$_4^-$ anion takes place, which stabilize the structure of **15** and **16**. Such an interaction is proposed in the calculated reaction mechanism of an yttrium bisborohydride species in the ROP of CL (Section 2.1, Scheme 19, C). For larger rare earth metal ions, increasing the size of the coordination sphere would decrease the interaction between the coordinated ligands and thereby reducing this stabilizing effect. In order to obtain more insight into the reaction mechanism, more experimental work is required. For example, the reaction of the lutetium chloro complex [(DIP$_2$pyr)LuCl$_2$(THF)$_2$][163] with sodium borohydride could probably lead to another product like [(DIP$_2$pyr)Lu(BH$_4$)$_2$(THF)$_n$], which would indicate that the described reactivity depends on the type of borohydride reagent. Possibly, the lutetium trisborohydride compared with sodium borohydride or the larger rare earth metal trisborohydrides is more active in attacking the Schiff-base function of the ligand. On the other hand, if this reaction would give the same product [{DIP$_2$pyr*-BH$_3$}Lu(BH$_4$)(THF)$_2$] (**16**), the reactivity might be a result of the (DIP$_2$pyr)⁻ ligand coordinated to different metal centers. However, more experimental and theoretical work needs to be done to prove these assumptions.

In addition to the chloride and the borohydride complexes described above, novel mixed borohydride-chloride complexes of the 2,5-bis{*N*-(2,6-diisopropylphenyl)iminomethyl}-pyrrolyl ligand were prepared.[203] As shown in Scheme 23, the reaction of [(DIP$_2$pyr)K] (**10**) with a 1: 1 mixture of the corresponding trisborohydrides [Ln(BH$_4$)$_3$(THF)$_3$] and anhydrous lanthanide trichlorides in THF at elevated temperature afforded the dimeric compounds [(DIP$_2$pyr)LnClBH$_4$(THF)]$_2$ (Ln = La (**17**), Nd (**18**)). As expected, the chlorine atoms act as bridging ligands, while the BH$_4$⁻ anions are η^3-coordinated. Mixed borohydride chloride complexes are unusual in rare earth metal chemistry and were only observed in clusters.[204, 205]

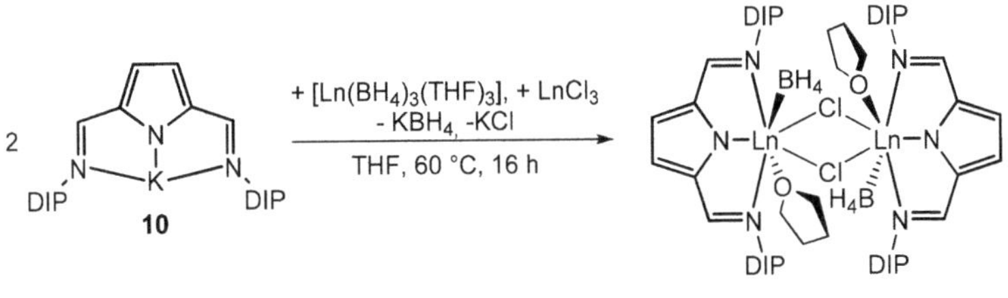

Scheme 23

Complexes **17** and **18** were characterized by standard analytical / spectroscopic techniques and the solid state structures were analyzed by single crystal X-ray diffraction. Both compounds are poorly soluble in THF and thus could not be investigated *via* NMR spectroscopy. The IR spectra of compounds **17** and **18** show two characteristic peaks at 2210 cm^{-1} and 2447 cm^{-1} (**17**) and 2229 cm^{-1} and 2451 cm^{-1} (**18**), which can be assigned to a terminally coordinated Ln(η^3–H$_3$B-H) unit.[178]

Compounds **17** and **18** are isostructural and crystallize in the monoclinic space group *C2/c* with four molecules of **17** or **18** and five molecules of THF in the unit cell. Figure 12 shows the solid state structure of 17 and selected bond lengths and angles are given for both compounds. The coordination polyhedron of each metal center in **17** and **18** is formed by the $(DIP_2pyr)^-$ ligand, two chlorine atoms, one BH_4^- anion and one THF molecule, showing a distorted pentagonal bipyramidal geometry. The THF molecule and the BH_4^- ligand form the apexes of the distorted bipyramid and exhibit an almost linear setup (O1-La-B 174.65(11)° (**17**) and O1-Nd-B 173.76(10)° (**18**)). The sum of the corresponding five valence angles (357.23° (**17**) and 358.03° (**18**)) of the LnN_3Cl_2 fragment shows an almost planar arrangement. In both compounds, the chlorine atoms are bridging the metal centers almost symmetrically (La-Cl 2.8695(10) Å, La-Cl′ 2.8733(10) Å (**17**) and Nd-Cl 2.7984(10) Å, Nd-Cl′ 2.8038(10) Å (**18**)). As observed for compound **12**, the $(DIP_2pyr)^-$ ligand is symmetrically attached to the metal center and the metal-imine nitrogen bond distances (La-N1 2.773(3) Å, La-N3 2.758(3) Å (**17**) and Nd-N1 2.736(3) Å, Nd-N3 2.720(3) Å (**18**)) are longer than the metal-pyrrolyl distance (La-N2 2.438(3) Å (**17**), Nd-N2 2.393(3) Å (**18**)). As observed for compound **14**, the BH_4^- groups show a η^3-coordination *via* the freely refined hydrogen atoms, which is typical for $Ln-BH_4$ compounds.[107, 110, 114, 115, 122, 178]

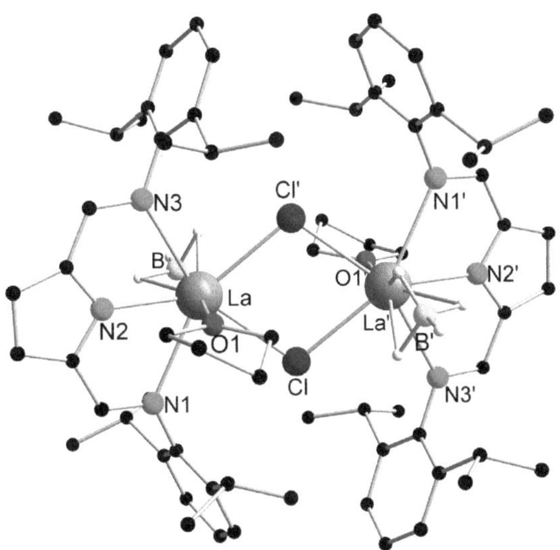

Figure 12 Solid state structure of **17**, omitting hydrogen atoms, except for the freely refined B-H atoms. Selected bond lengths [Å] or angles [°] are given for **17** and isostructural complex **18**.

17:La-N1 2.773(3), La-N2 2.438(3), La-N3 2.758(3), La-O1 2.554(2), La-B 2.682(4), La-Cl 2.8695(10), La-Cl′ 2.8733(10); N1-La-N2 61.73(8), N1-La-N3 123.17(8), N1-La-O1 83.45(8), N1-La-B 93.83(12), N1-La-Cl 80.15(6), N1-La-Cl′ 152.63(5), N2-La-N3 61.44(8), N2-La-O1 75.95(8), N2-La-B 98.70(12), N2-La-Cl 138.94(6), N2-La-Cl′ 136.48(6), N3-La-O1 83.75(8), N3-La-B 93.95(12), N3-La-Cl 152.79(6), N3-La-Cl′ 78.81(6), O1-La-B 174.65(11), O1-La-Cl 85.61(6), O1-La-Cl′ 83.19(6), B-La-Cl 98.49(10), B-La-Cl′ 101.14(10), Cl-La-Cl′ 75.10(3), La-Cl-La′ 104.90(3).

18: Nd-N1 2.736(3), Nd-N2 2.393(3), Nd-N3 2.720(3), Nd-O1 2.496(2), Nd-B 2.612(4), Nd-Cl 2.7984(10), Nd-Cl′ 2.8038(10); Nd-Cl-Nd′ 105.02(4), N1-Nd-N2 62.58(8), N1-Nd-N3 125.00(8), N1-Nd-O1 84.26(8), N1-Nd-B 92.98(12), N1-Nd-Cl 79.56(6), N1-Nd-Cl′ 152.57(6), N2-Nd-N3 62.45(9), N2-Nd-O1 76.07(8), N2-Nd-B 97.69(11), N2-Nd-Cl 139.42(7), N2-Nd-Cl′ 137.30(7), N3-Nd-O1 84.63(9), N3-Nd-B 92.43(12), N3-Nd-Cl 152.72(6), N3-Nd-Cl′ 78.46(6), O1-Nd-B 173.76(10), O1-Nd-Cl 86.67(6), O1-Nd-Cl′ 84.24(6), B-Nd-Cl 98.35(10), B-Nd-Cl′ 100.58(10), Cl-Nd-Cl′ 74.98(4).

Catalytic application

Since neodymium derivatives play an important role as Ziegler-Natta catalysts for industrial polymerization, the (DIP$_2$pyr)$^-$ chloride and borohydride complexes of neodymium, [(DIP$_2$pyr)NdCl$_2$(THF)]$_2$ (**12**) and [(DIP$_2$pyr)Nd(BH$_4$)$_2$(THF)$_2$] (**14**) were investigated for the polymerization of 1,3-butadiene to yield poly-*cis*-1,4-butadiene.[196] The polymerization studies in the presence of various cocatalysts and cyclohexane as solvent at 65 °C were performed by S. K.-H. Thiele. The results of the polymerization experiments are shown in Table 3 and Table 4. In all reactions, the catalyst concentration was in the range of 7.4 - 7.6 x 10^{-5} mol/l and the molar ratio of 1,3-butadiene to the catalysts was in the range of about 20000 to 22600. For those runs with high butadiene conversions, the micro structure of the resulting polymer was established. The observed activity strongly depends on the nature of the cocatalyst. Table 3 shows the results for the polymerization of 1,3-butadiene by using modified methylalumoxane, comprised of trimethylaluminum and triisobutylaluminum (MMAO-3A; 7 wt % solution in heptane (Akzo Nobel)) as cocatalyst. In these reactions, the chloride compound **12** is more active than the borohydride **14** and shows a *cis*-selectivity of 85 %. In addition, the polymerization of 1,3-butadiene was investigated by using either a mixture of AlEt$_3$ / B(C$_6$F$_5$)$_3$ or of AlEt$_3$ / [PhNMe$_2$H][B(C$_6$F$_5$)$_4$] as cocatalysts (Table 4). Again the highest turnover frequency was observed for compound **12**, by using AlEt$_3$ / B(C$_6$F$_5$)$_3$ (7.2 mmol: 100.98 µmol) as cocatalyst mixture (Table 4, entry 3). The observed activity of 3.606 [kg of polymer /(mmol catalyst x h)] based on the 10 min result is significantly higher compared to other systems tested under similar conditions such as lithium Nd(versatate)$_3$/MAO (1:264)[187] or (hexa-1,5-diene-1,6-diamide)neodymium dibromide.[192] The highest *cis* selectivitiy of about 85 % was observed for compound **12**, by using AlEt$_3$ / [PhNMe$_2$H][B(C$_6$F$_5$)$_4$] (7.2 mmol: 99.4 µmol) as cocatalyst (Table 4, entry 4). The selectivities observed in the presented studies are lower than those obtained by the (bis(*N*-2,6-dialkylphenyl)isophthalaldimine-2-yl) complexes (99.3-99.9 %) which were recently published.[195]

Table 3 1,3-Butadiene polymerization catalyzed by **12** and **14** and MMAO as cocatalyst.[a][b]

entry	1	2
catalyst	**14**	**12**
catalyst [μmol]	48.18	48.18
MMAO [mmol]	14	14
amount of 1,3-butadiene [mol]	1.01	1.09
1,3-butadiene/Nd molar ratio	20.986	22.597
butadiene conversion (15 min) [%]	8.5	42.2
butadiene conversion (1h) [%]	10.8	84.9
activity [kg polymer /(mmol Nd x hour)], based on 15 min result	0.176	0.876
cis-1,4-PBR [%]	n.d.	84.9
trans-1,4-PBR [%]	n.d.	13.3
1,2-PBR [%]	n.d.	1.8
M_w [g/mol]	680000	577000
M_n [g/mol]	50000	247000
D	10	2.3
T_g / °C	-107.9	-107.4

[a] The polymerization experiments were performed at Dow Olefinverbund GmbH. [b] reaction temperature 65 °C, Solvent: cyclohexane: 640-650 ml.

Table 4 1,3-Butadiene polymerization catalyzed by **12** and **14** and cocatalyst mixtures.[a][b]

entry	1	2	3	4
catalyst	**14**	**14**	**12**	**12**
catalyst [µmol]	50.1	49.4	50.4	49.5
$AlEt_3$ [mmol]	7.2	7.2	7.2	7.2
$B(C_6F_5)_3$ [µmol]	101.34	-	100.98	-
$[PhNMe_2H][B(C_6F_5)_4]$ [µmol]	-	99.4	-	99.4
amount of 1,3-butadiene [mol]	1.05	1.09	1.10	1.01
1,3-butadiene/Nd molar ratio	20.878	22.105	19.980	20.343
butadiene conversion (10 min) [%]	27.5	35.6	51.4	18.5
butadiene conversion (1h) [%]	44.9	85.7	94.9	19.4
activity [kg polymer /(mmol Nd x hour)], based on 10 min result	1.801	2.365	3.604	1.321
cis-1,4-PBR [%]	83.9	75.2	76.1	85.3
trans-1,4-PBR [%]	13.1	23.3	23.1	13.7
1,2-PBR [%]	3	1.5	0.8	1
M_w [g/mol]	418000	322500	673500	n.d.
M_n [g/mol]	78000	112000	246000	n.d.
D	5.37	2.88	2.74	-
T_g / °C	-107.0	-104.8	-106.1	-106.6

[a] The polymerization experiments were performed at Dow Olefinverbund GmbH. [b] reaction temperature 65 °C, Solvent: cyclohexane: 640 -650 ml.

In summary, 2,5-bis{*N*-(2,6-diisopropylphenyl)iminomethyl}pyrrolyl rare earth metal chlorides and borohydrides, including the novel mixed borohydride chloride complexes were successfully prepared. The chloride $[(DIP_2pyr)NdCl_2(THF)]_2$, the borohydrides $[(DIP_2pyr)Ln(BH_4)_2(THF)_2]$ (Ln = La, Nd) and the mixed borohydride chloride complexes $[(DIP_2pyr)LnClBH_4(THF)]_2$ (Ln = La, Nd) were the first examples of rare earth metals with larger ionic radii, coordinating the 2,5-bis{*N*-(2,6-diisopropylphenyl)iminomethyl}pyrrolyl ligand. The borohydrides of the larger rare earth metals exhibit the expected composition $[(DIP_2pyr)Ln(BH_4)_2(THF)_2]$ (Ln = La, Nd), while for the smaller rare earth metals an unusual redox reaction of a BH_4^- anion with one of the Schiff-base functions of the ligand was observed, leading to the products $[\{DIP_2pyr^*-BH_3\}Ln(BH_4)(THF)_2]$ (Ln = Sc, Lu). The reason for the different reactivity is unclear. In addition, the two neodymium containing borohydride and chloride complexes were investigated as Ziegler-Natta catalysts for the polymerization of 1,3-butadiene to form poly-*cis*-1,4-butadiene, by using various cocatalyst mixtures. In particular, by using the chloride complex, very high activities and good selectivities were achieved.

3.2.3 2,5-Bis{*N*-(2,6-diisopropylphenyl)iminomethyl}pyrrolyl complexes of divalent lanthanides and alkaline earth metals

Since the 2,5-bis{*N*-(2,6-diisopropylphenyl)iminomethyl}pyrrolyl ligand has been shown to stabilize trivalent rare earth metal complexes,[158, 163, 164, 200] we were interested in introducing this ligand into the chemistry of divalent lanthanides. Similarly to alkaline earth metals, divalent lanthanides tend to undergo Schlenk-like redistribution reactions. For example, in the synthesis of $Ln(Dpp)I(THF)_n$ (Ln = Eu, Yb; Dpp = $2,6\text{-}Ph_2C_6H_3$), the redistribution products $Ln(Dpp)_2(THF)_2$ and $LnI_2(THF)_n$ were observed.[206] By using ytterbium as the center metal, the shift of the Schlenk-like equilibrium could be controlled by the different solubility of the species, while for europium only the homoleptic complex $Eu(Dpp)_2(THF)_2$ was obtained. On the contrary, the bis(phosphinimino)methanide iodo complexes of the divalent lanthanides synthesized by our group were obtained as heteroleptic complexes.[146, 148] While the ytterbium complex $[\{CH(PPh_2NSiMe_3)_2\}YbI(THF)_2]$ is monomeric in the solid state, the complexes of samarium and europium $[\{CH(PPh_2NSiMe_3)_2\}LnI(THF)]_2$ are dimeric. In regard to these observations, divalent lanthanide complexes of the 2,5-bis{*N*-(2,6-diisopropylphenyl)iminomethyl}pyrrolyl ligand were synthesized. In addition, it is well established that the reactivity and coordination behavior of the divalent lanthanides and the heavier alkaline earth metals are related, which is a result of the similar ionic radii (Section 1.1 and 1.2). Therefore, it is evident to study the coordination behavior of the 2,5-bis{*N*-(2,6-diisopropylphenyl)iminomethyl}pyrrolyl ligand in alkaline earth metal chemistry. Since only few alkaline earth metal complexes were investigated towards their catalytic activity in hydroamination reactions (Section 1.2), potential catalytically active species were formed by introducing the $\{N(SiMe_3)_2\}^-$ ligand as a leaving group. The resulting compounds were investigated for the hydroamination of aminoalkenes and aminoalkynes.

As shown in Scheme 24, the reaction of $[(DIP_2pyr)K]$ (**10**) with anhydrous lanthanide diiodides in THF at elevated temperature afforded the heteroleptic complexes $[(DIP_2pyr)LnI(THF)_3]$ (Ln = Sm (**19**), Eu (**20**), Yb (**21**)). Independently from the ionic radii of the center metals, all complexes are monomeric in the solid state with three THF molecules coordinated to saturate the coordination sphere, and the $(DIP_2pyr)^-$ ligands show an η^3-coordination mode.

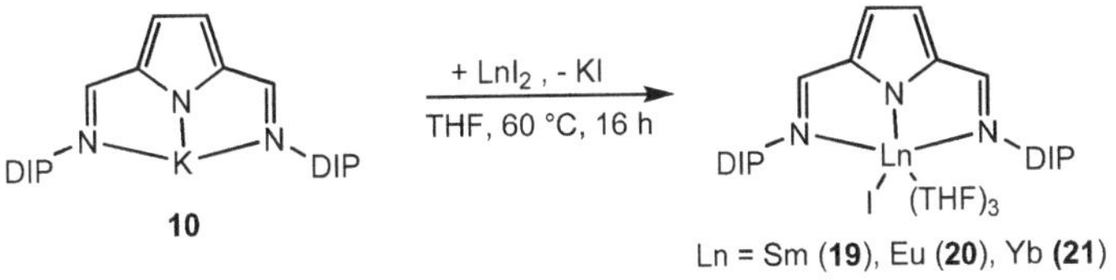

Scheme 24

Reaction of **10** with LnI_2 ($+ LnI_2$, $- KI$; THF, 60 °C, 16 h) to give products with Ln = Sm (**19**), Eu (**20**), Yb (**21**).

The novel complexes **19-21** were characterized by standard analytical / spectroscopic techniques and the solid state structures were analyzed by single crystal X-ray diffraction. The NMR spectra of the diamagnetic compound **21** show the expected signals for the $(DIP_2pyr)^-$ ligand. In the ^1H NMR spectrum, one doublet (δ = 1.25 ppm) and one septet (δ = 3.25 ppm) observed for the isopropyl groups indicate that the 2,6-diisopropylaniline moieties of the ligand can freely rotate in solution. The $^{13}C\{^1H\}$ NMR data are consistent with these observations. The $(DIP_2pyr)^-$ ligand in **21** is asymmetrically coordinated in the solid state and the structure will be discussed later in this section. In the ^1H NMR spectrum of **21**, the singlet at δ = 8.21 ppm for the imino N=CH units indicates an η^3-coordination mode. The NMR data indicate the same chemical environment for the isopropyl groups and the N=CH moieties in solution.

Crystals of the samarium and europium derivatives **19** and **20** were obtained from hot THF as black (**19**) and red (**20**) rods. Compounds **19** and **20** are isostructural in the solid state and crystallize in the monoclinic space group $P2_1/c$ with four molecules of the corresponding complex and eight THF molecules in the unit cell. The solid state structure of **19** is illustrated in Figure 13 and selected bond lengths and angles are given for compounds **19** and **20**. Surprisingly, complexes **19** and **20** are monomeric in the solid state and the coordination sphere is satisfied by three additionally coordinated THF molecules. The $(DIP_2pyr)^-$ ligands are coordinated almost symmetrically in an η^3-mode to the metal center (Sm-N1 2.922(6) Å, Sm-N2 2.473(5) Å, Sm-N3 2.937(5) Å (**19**) and Eu-N1 2.918(7) Å, Eu-N2 2.436(7) Å, Eu-N3 2.955(7) Å (**20**)). The coordination polyhedron of each metal center in **19** and **20** is formed by the $(DIP_2pyr)^-$ ligand, the iodine atom and three THF molecules and shows a distorted pentagonal bipyramidal geometry. The apexes of each bipyramid are formed by the iodine atom and one THF molecule with an almost linear setup (O1-Sm-I 171.35(12)° (**19**) and O1A-Eu-I 172.8(3)° (**20**)). The nearly planar arrangement of the LnN_3O_2 fragments is confirmed by the sums of the corresponding five valence angles (358.1° (**19**) and 358.4° (**20**)). Interestingly, the iodine atoms occupy a *cisoid* position towards all three nitrogen atoms of the $(DIP_2pyr)^-$ ligands (N1-Sm-I 99.03(11)°,

N2-Sm-I 104.96(14)°, N3-Sm-I 98.83(11)° (**19**) and N1-Eu-I 98.55(2)°, N2-Eu-I 104.4(2)°, N3-Eu-I 98.28(15)° (**20**)). The Ln-I bond distances (Sm-I 3.2230(9) Å (**19**) and Eu-I 3.1926(11) Å (**20**)) are similar to those observed for monomeric samarium(II) and europium(II) iodo complexes stabilized by bulky ligands.[207, 208]

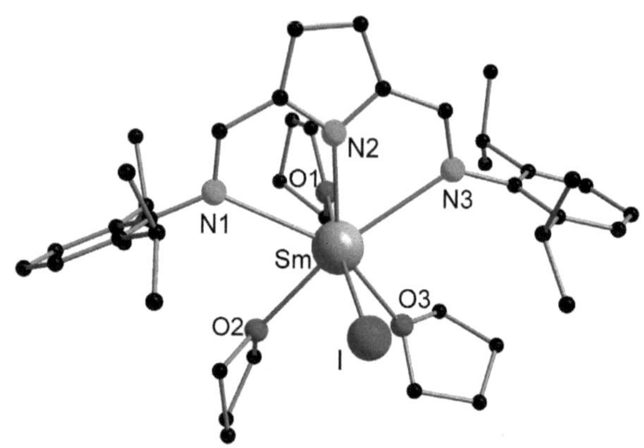

Figure 13 Solid-state structure of **19**, omitting hydrogen atoms. Selected bond lengths [Å] or angles [°] are given for **19** and the isostructural complex **20**.

19: Sm-N1 2.922(6), Sm-N2 2.473(5), Sm-N3 2.937(5), Sm-O1 2.592(5), Sm-O2 2.600(5), Sm-O3 2.582(5), Sm-I 3.2230(9); N1-Sm-N2 61.2(2), N1-Sm-N3 121.93(15), N2-Sm-N3 60.9(2), N1-Sm-O1 89.1(2), N1-Sm-O2 75.6(2), N1-Sm-O3 151.4(2), N2-Sm-O1 81.4(2), N2-Sm-O2 135.6(2), N2-Sm-O3 143.3(2), N3-Sm-O1 79.04(15), N3-Sm-O2 157.9(2), N3-Sm-O3 83.6(2), N1-Sm-I 99.03(11), N2-Sm-I 104.96(14), N3-Sm-I 98.83(11), O1-Sm-I 171.35(12), O2-Sm-I 90.65(11), O3-Sm-I 88.35(11), O1-Sm-O2 88.61(15), O2-Sm-O3 76.8(2), O1-Sm-O3 83.1(2).

20 (The THF molecule around O1 is disordered and selected bond lengths [Å] and angles [°] are given for both positions O1A and O1B): Eu-N1 2.918(7), Eu-N2 2.436(7), Eu-N3 2.955(7), Eu-O1A 2.648(14), Eu-O1B 2.293(13), Eu-O2 2.582(8), Eu-O3 2.564(7), Eu-I 3.1926(11); N1-Eu-N2 61.4(2), N1-Eu-N3 122.3(2), N2-Eu-N3 61.0(2), N1-Eu-O1A 87.9(3), N1-Eu-O1B 93.1(4), N1-Eu-O2 75.8(2), N1-Eu-O3 152.7(2), N2-Eu-O1A 81.4(3), N2-Eu-O1B 87.7(4), N2-Eu-O2 136.3(2), N2-Eu-O3 142.6(2), N3-Eu-O1A 80.7(3), N3-Eu-O1B 81.5(4), N3-Eu-O2 158.3(2), N3-Eu-O3 82.5(2), N1-Eu-I 98.55(2), N2-Eu-I 104.4(2), N3-Eu-I 98.28(15), O1A-Eu-I 172.8(3), O1B-Eu-I 166.2(3), O2-Eu-I 89.8(2), O3-Eu-I 87.7(2), O1A-Eu-O2 88.8(3), O1B-Eu-O2 85.9(4), O2-Eu-O3 77.7(2), O1A-Eu-O3 85.1(3), O1B-Eu-O3 78.5(4).

Crystallization from THF / *n*-pentane afforded **21** as green plates in the monoclinic space group $P2_1/c$ with four molecules of the corresponding complex and two *n*-pentane molecules in the unit cell. The solid state structure of **21**, shown in Figure 14, is very similar to that of **19** and **20**, and the coordination polyhedron of **21** exhibits the same distorted pentagonal bipyramidal geometry. Differently from **19** and **20**, the apexes of the bipyramid are formed by two THF molecules, which are arranged almost linearly (O1-Yb-O3 170.3(5)° (**21**)). The sum of the corresponding five valence angles shows a nearly planar arrangement of the LnN$_3$OI fragment (359.9°). In contrast to **19** and **20**, the (DIP$_2$pyr)⁻ ligand in **21** is coordinated asymmetrically to the metal center and one Schiff-base function (N3) shows an Yb-N bond distance of 2.751(13) Å, while the other one exhibits a long interaction to the metal center (Yb-N1 3.157(14) Å). The bond length of the pyrrolyl unit (Yb-N2 2.400(10) Å) is slightly smaller than those observed for **19** and **20**. The asymmetrical coordination of the (DIP$_2$pyr)⁻ ligand is certainly caused by the smaller ionic radius of the metal center in **21** compared with those of **19** and **20**, and could be additionally a result of the crystal packing (see analogous calcium compound **24** later in this section). The Yb-I bond distance (3.100(2) Å) of **21** is within the same range observed for other monomeric ytterbium (II) iodo complexes.[146, 206, 209]

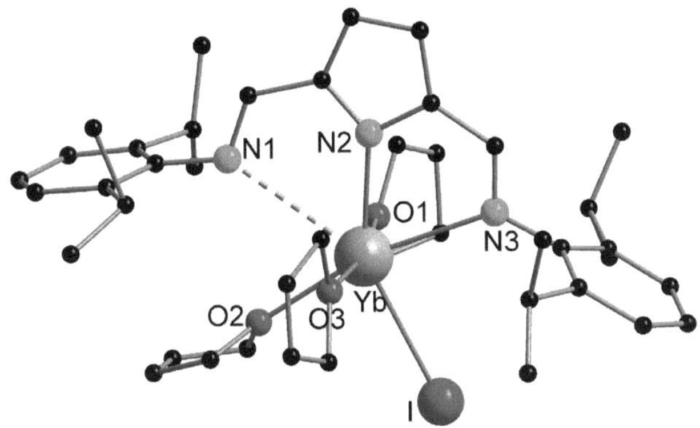

Figure 14 Solid-state structure of **21**, omitting hydrogen atoms. Selected bond lengths [Å] or angles [°]: Yb-N1 3.157(14), Yb-N2 2.400(10), Yb-N3 2.751(13), Yb-O1 2.471(14), Yb-O2 2.453(11), Yb-O3 2.433(10), Yb-I 3.100(2); N1-Yb-N2 58.6(4), N1-Yb-N3 123.8(4), N2-Yb-N3 65.3(4), N1-Yb-O1 87.3(4), N1-Yb-O2 68.3(4), N1-Yb-O3 83.8(4), N2-Yb-O1 90.4(4), N2-Yb-O2 127.0(5), N2-Yb-O3 88.4(4), N3-Yb-O1 96.3(4), N3-Yb-O2 167.5(4), N3-Yb-O3 92.0(4), N1-Yb-I 152.7(3), N2-Yb-I 148.6(4), N3-Yb-I 83.3(3), O1-Yb-I 93.6(4), O2-Yb-I 84.4(3), O3-Yb-I 92.4(3), O1-Yb-O2 86.7(5), O2-Yb-O3 86.2(4), O1-Yb-O3 170.3(5).

Interestingly, all complexes **19-21** are monomeric in the solid state, which is unusual, since samarium and europium with larger ionic radii tend to form dimeric complexes bridged by iodine atoms.[146, 148, 210-216] In the literature, there are only few examples for monomeric heteroleptic iodo complexes of divalent samarium and europium. For samarium(II), there is one example with a very bulky ligand {$C_5Me_4SiMe_2(iPr_2$-tacn)}$^-$ (tacn = 1,4-diisopropyl-1,4,7-triazacyclononane).[208] This ligand coordinates *via* the cyclopentadienyl moiety and the three nitrogen atoms of the triazacyclononane ring to the samarium(II) center. Consequently, the coordination sphere is already saturated and a monomer is formed. To the best of my knowledge, the only example for a monomeric heteroleptic iodo complex of europium(II) is the cluster [$IEu(OtBu)_4${Li(THF)}$_4$(OH)].[207] This compound is stabilized by a bulky alkali metal cage on one side and thus cannot be compared to **19** and **20**. The ($DIP_2pyr)^-$ ligand is not as sterically crowded and the coordination sphere of **19** and **20** is satisfied by three additionally coordinated THF molecules. In the literature, the only example for such a structure is the samarium(II) compound [$Sm(L^{Ph,tBu})I(THF)_4$] ($L^{Ph,tBu}$ = {$N(C(Ph)=N)_2C(tBu)Ph$}$^-$), which coordinates four THF molecules to saturate the coordination sphere instead of forming a dimer.[217] In addition, the nitrogen atom of the monodentate ($L^{Ph,tBu})^-$ ligand and the iodine atom exhibit a relatively small angle (N1-Sm-I 100.25(9)°) instead of building a nearly linear N-Sm-I unit. This arrangement was also observed for **19** and **20**, as described above. The reason for the formation of the *cisoid* isomers is unclear. The crystal structures of the diodides of samarium(II) and europium(II) coordinated by different solvents are comprehensively described in the literature and *transoid* as well as *cisoid* isomers were observed.[218-223] The formation of different isomers seems to be not only an effect of coordinated solvents, but also of the crystallization temperatures and the crystal packing. Probably, more information of the structural motif of **19** and **20** could be obtained by performing computational studies.

Furthermore, the coordination chemistry of the 2,5-bis{N-(2,6-diisopropylphenyl)imino-methyl}pyrrolyl ligand was expanded to the alkaline earth metals. The corresponding complexes [(DIP$_2$pyr)MI(THF)$_n$] (M = Ca (**24**), Sr (**22**) (n = 3); Ba (**23**) (n = 4)) were synthesized by the reaction of [(DIP$_2$pyr)K] (**10**) with anhydrous alkaline earth metal diiodides in THF at elevated temperature. The strontium and barium complexes **22** and **23** were obtained after 16 h (Scheme 25), while for the synthesis of the calcium compound **24** an elongated reaction time of 60 h was required (Scheme 26).

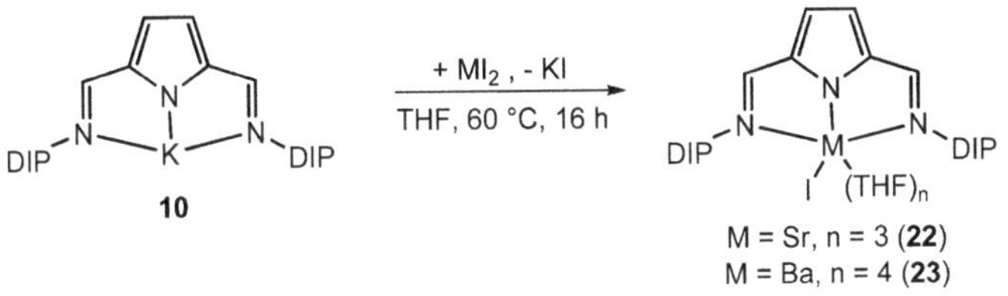

Scheme 25

As observed for the divalent lanthanide analogues, **22-24** are monomeric in the solid state. Interestingly, the (DIP$_2$pyr)⁻ ligand exhibits different coordination modes in relation to the ionic radii of the center metals. While the (DIP$_2$pyr)⁻ ligands in **22** and **23** coordinate in an η^3-fashion to the metal centers (Scheme 25), the calcium complex **24** shows an η^2-coordinated ligand (Scheme 26). This is remarkable, since the analogous ytterbium complex **21** shows an η^3-coordination mode of the ligand which is coordinated asymmetrically, but by all three nitrogen atoms to the metal center.

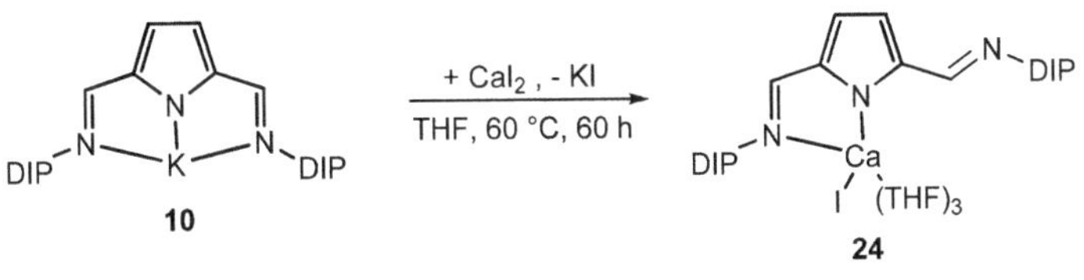

Scheme 26

Complexes **22-24** were characterized by standard analytical / spectroscopic techniques and the solid state structures were analyzed by single crystal X-ray diffraction. In the ^{1}H NMR spectra of **22** and **23** a multiplet (δ = 1.17-1.21 (**22**), 1.16-1.32 (**23**) ppm) and a clear septet (δ = 3.22 (**22**), 3.22 (**23**) ppm) are observed for the isopropyl groups of the (DIP$_2$pyr)⁻ ligands, indicating a restricted rotation of the 2,6-diisopropylaniline moieties. The sharp singlet for the imino N=CH units (δ = 8.01 (**22**), 8.03 (**23**) ppm) confirms the symmetrical coordination mode of the ligands in **22** and **23**. On the contrary, the ^{1}H NMR spectrum of **24** shows one doublet (δ = 1.20 ppm) and one septet (δ = 3.22 ppm) for the isopropyl groups, indicating that the 2,6-diisopropylaniline moieties of the ligand can freely rotate in solution. For the imino N=CH moieties a broad singlet is observed (δ = 8.24 ppm). These observations indicate that both 2,6-diisopropylaniline moieties of **24** exhibit a similar chemical environment in solution at room temperature. In order to investigate the coordination mode of the (DIP$_2$pyr)$^-$ ligand in **24** in solution, further VT-^{1}H NMR studies were performed. At 65 °C the ^{1}H NMR spectrum is very similar to the one obtained at room temperature, except for the imino N=CH moieties which show a clear singlet (δ = 8.23 ppm) caused by the thermal motion. However, at -70 °C the ^{1}H NMR spectrum shows a multiplet for the isopropyl CH$_3$ groups (δ = 1.09-1.28). This indicates a restricted rotation of the 2,6-diisopropylaniline moieties of the ligand, which was expected at low temperature. The pyrrolyl hydrogen atoms as well as the imino N=CH moieties show a clear singlet (δ = 6.59 ppm and δ = 8.07 ppm), which indicates a similar chemical environment for both 2,6-diisopropylaniline moieties of the ligand. In solution even at a low temperature, there is no evidence of an asymmetrical coordination mode of the (DIP$_2$pyr)$^-$ ligand.

As expected, **22** is isostructural to the samarium and the europium compounds **19** and **20** and crystallizes in the monoclinic space group $P2_1/c$ with four molecules of the complex and eight THF molecules in the unit cell. As shown in Figure 15, the (DIP$_2$pyr)$^-$ ligand is coordinated almost symmetrically in an η^3-mode to the metal center (Sr-N1 2.969(5) Å, Sr-N2 2.479(5) Å, Sr-N3 2.988(5) Å). The Sr-I bond distance (3.2262(9) Å) is similar to the Ln-I bond lengths observed for **19** and **20** (Sm-I 3.2230(9) Å (**19**) and Eu-I 3.1926(11) Å (**20**)). Similarly to **19** and **20**, the coordination sphere of the metal center of **22** forms a distorted pentagonal bipyramid with an almost planar arrangement of the SrN$_3$O$_2$ fragment, which is confirmed by the sum of the corresponding five valence angles (358.4°). The apexes of the distorted pentagonal bipyramid, formed by the iodine atom and one THF molecule show a nearly linear setup (O1-Sr-I 172.05(12)°). As observed for **19** and **20**, the iodine atom is *cisoid* to all three nitrogen atoms of the (DIP$_2$pyr)$^-$ ligand (N1-Sr-I 98.17(10)°, N2-Sr-I 104.24(11)°, N3-Sr-I 98.88(9)°).

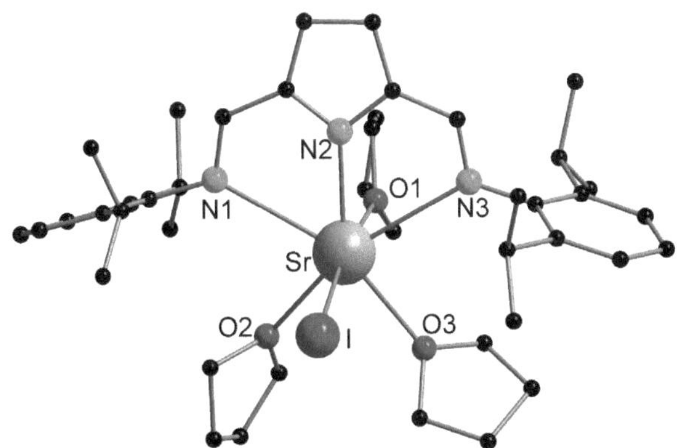

Figure 15 Solid-state structure of **22**, omitting hydrogen atoms. Selected bond lengths [Å] or angles [°]: Sr-N1 2.969(5), Sr-N2 2.479(5), Sr-N3 2.988(5), Sr-O1 2.570(5), Sr-O2 2.575(4), Sr-O3 2.574(4), Sr-I 3.2262(9); N1-Sr-N2 61.20(14), N1-Sr-N3 121.66(13), N2-Sr-N3 60.58(14), N1-Sr-O1 88.8(2), N1-Sr-O2 76.00(14), N1-Sr-O3 152.67(15), N2-Sr-O1 82.5(2), N2-Sr-O2 136.3(2), N2-Sr-O3 142.85(15), N3-Sr-O1 88.5(2), N3-Sr-O2 158.70(14), N3-Sr-O3 83.14(14), N1-Sr-I 98.17(10), N2-Sr-I 104.24(11), N3-Sr-I 98.88(9), O1-Sr-I 172.05(12), O2-Sr-I 89.31(12), O3-Sr-I 88.03(11), O1-Sr-O2 88.7(2), O2-Sr-O3 77.5(2), O1-Sr-O3 84.02(15).

Compound **23** crystallizes in the monoclinic space group $P2_1$ with four molecules of the complex and four THF molecules in the unit cell. The asymmetric unit contains two independent molecules of **23** and two THF molecules. The (DIP$_2$pyr)⁻ ligand is coordinated almost symmetrically in an η^3-mode to the metal center (Figure 16). Due to the larger ionic radius of the center metal in **23**, the Ba-N bond distances are longer (Ba1-N1 3.196(4) Å, Ba1-N2 2.690(3) Å, Ba1-N3 3.145(3) Å) than those observed for **22**. Similarly to **19**, **20** and **22**, complex **23** is monomeric in the solid state and the coordination sphere of the metal center is satisfied by four additionally coordinated THF molecules, instead of forming a dimer. Therefore, **23** shows an eight-fold coordination sphere formed by the (DIP$_2$pyr)⁻ ligand, the iodine atom and four THF molecules. The iodine atom of **23** is not arranged linearly with respect to any of the nitrogen atoms of the (DIP$_2$pyr)⁻ ligand (N1-Ba1-I1 128.59(6)°, N2-Ba1-I1 148.40(7)°, N3-Ba1-I1 106.04(7)°), but the angle between the pyrrolyl unit and the iodine atom is larger than those observed for **19**, **20** and **22**. The Ba-I bond distance of (Ba1-I1 3.4950(8) Å) is similar to that observed for the monomeric barium iodo complex stabilized by the very bulky tris[3-(2-methoxy-1,1-dimethyl)pyrazolyl]hydroborate (TpC*) ligand.[224]

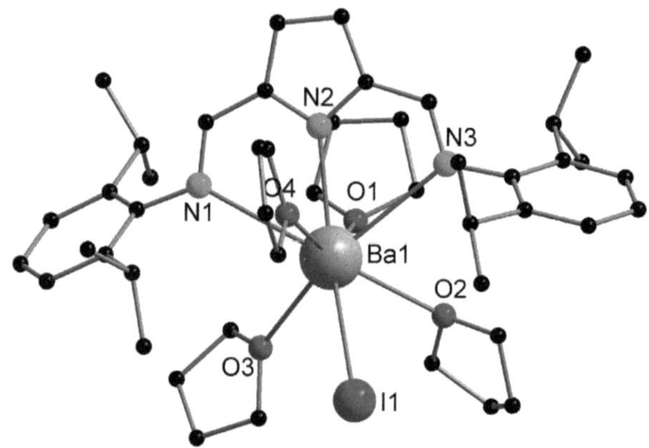

Figure 16 Solid-state structure of **23**, omitting hydrogen atoms. Only one of the two independent molecules is shown. Selected bond lengths [Å] or angles [°]: Ba1-N1 3.196(4), Ba1-N2 2.690(3), Ba1-N3 3.145(3), Ba1-O1 2.812(3), Ba1-O2 2.801(3), Ba1-O3 2.806(3), Ba1-O4 2.877(3), Ba1-I1 3.4950(8), Ba2-N4 3.214(4), Ba2-N5 2.679(3), Ba2-N6 3.156(3), Ba2-O5 2.819(3), Ba2-O6 2.774(4), Ba2-O7 2.865(3), Ba2-O8 2.776(3), Ba2-I2 3.4922(7); N1-Ba1-N2 57.13(10), N1-Ba1-N3 115.04(9), N2-Ba1-N3 57.94(10), N1-Ba1-O1 79.94(9), N1-Ba1-O2 147.02(10), N1-Ba1-O3 74.46(10), N1-Ba1-O4 77.66(10), N2-Ba1-O1 74.37(9), N2-Ba1-O2 119.27(10), N2-Ba1-O3 126.22(11), N2-Ba1-O4 71.75(9), N3-Ba1-O1 84.92(9), N3-Ba1-O2 72.28(10), N3-Ba1-O3 156.64(9), N3-Ba1-O4 81.52(10), N1-Ba1-I1 128.59(6), N2-Ba1-I1 148.40(7), N3-Ba1-I1 106.04(7), O1-Ba1-I1 134.84(7), O2-Ba1-I1 73.56(7), O3-Ba1-I1 79.99(8), O4-Ba1-I1 79.31(7), O1-Ba1-O2 68.43(10), O1-Ba1-O3 75.58(10), O1-Ba1-O4 145.75(9), O2-Ba1-O3 88.44(11), O2-Ba1-O4 134.71(11), O3-Ba1-O4 121.82(10) N4-Ba2-N5 56.92(9), N4-Ba2-N6 115.20(9), N5-Ba2-N6 58.28(10), N4-Ba2-O5 85.62(9), N4-Ba2-O6 159.73(11), N4-Ba2-O7 79.68(10), N4-Ba2-O8 73.89(10), N5-Ba2-O5 74.66(9), N5-Ba2-O6 126.08(10), N5-Ba2-O7 147.29(9), N5-Ba2-O8 118.65(11), N6-Ba2-O5 78.57(9), N6-Ba2-O6 71.77(10), N6-Ba2-O7 81.55(10), N6-Ba2-O8 143.67(9), N4-Ba2-I2 106.90(6), N5-Ba2-I2 149.15(7), N6-Ba2-I2 128.84(7), O5-Ba2-I2 133.86(6), O6-Ba2-I2 78.85(8), O7-Ba2-I2 78.65(7), O8-Ba2-I2 74.35(7), O5-Ba2-O6 77.00(11), O5-Ba2-O7 147.29(9), O5-Ba2-O8 66.78(10), O6-Ba2-O7 120.59(12), O6-Ba2-O8 89.53(12), O7-Ba2-O8 134.24(10).

Compound **24** crystallizes in the monoclinic space group $P2_1/c$ with four molecules of the complex and four THF molecules in the unit cell. Differently from the analogous ytterbium derivative **21**, the (DIP$_2$pyr)$^-$ ligand of **24** exhibits an η^2-coordination mode in the solid state (Figure 17). The ligand is coordinated by one Schiff-base function and the pyrrolyl moiety (Ca-N1 2.510(5) Å and Ca-N2 2.408(5) Å). As expected, the coordinated imino N=CH unit exhibits a slightly larger N-C bond length (N1-C1 1.296(8) Å) than the non coordinated N=CH moiety (N3-C6 1.262(7) Å). Similarly to **21**, the (DIP$_2$pyr)$^-$ ligand, the iodine atom and three THF molecules are coordinated to the metal center. Since in **24** only two coordination sites of the metal center are occupied by the (DIP$_2$pyr)⁻ ligand, a six-fold coordination sphere is formed, showing a distorted octahedral geometry. The square area of the distorted octahedron is formed by the pyrrolyl moiety and the three THF molecules. The almost planar arrangement of the CaO$_3$N fragment is confirmed by the sum of the corresponding four valence angles (360.2°). The coordinated Schiff-base moiety and the iodine atom form the apexes of the distorted octahedron and exhibit a nearly linear setup (N1-Ca-I 178.34(12)°), which is different from the ytterbium complex **21** (N1-Yb-I 152.7(3)°). The Ca-I bond distance (3.0782(13) Å) observed for **24** is similar to the Yb-I bond length (3.100(2) Å) of **21** and to [{(Me$_3$SiNPPh$_2$)$_2$CH}CaI(THF)$_2$] or [{(*i*Pr)$_2$ATI}CaI(THF)$_3$] ((*i*Pr)$_2$ATI = *N*-iso-propyl-2-(isopropylamino)troponiminate), synthesized by our group.[146, 225]

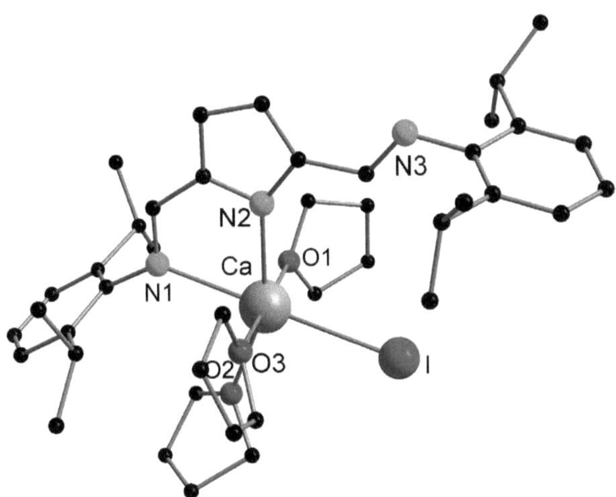

Figure 17 Solid-state structure of **24**, omitting hydrogen atoms. Selected bond lengths [Å] or angles [°]: Ca-N1 2.510(5), Ca-N2 2.408(5), Ca-O1 2.367(4), Ca-O2 2.371(5), Ca-O3 2.368(4), Ca-I 3.0782(13), N1-C1 1.296(8), N3-C6 1.262(7); N1-Ca-N2 71.6(2), N1-Ca-O1 90.31(15), N1-Ca-O2 87.8(2), N1-Ca-O3 89.74(15), N2-Ca-O1 92.0(2), N2-Ca-O2 159.4(2), N2-Ca-O3 85.13(15), N1-Ca-I 178.34(12), N2-Ca-I 107.21(11), O1-Ca-I 90.84(11), O2-Ca-I 93.41(13), O3-Ca-I 89.05(11), O1-Ca-O2 88.0(2), O2-Ca-O3 95.1(2), O1-Ca-O3 176.9(2).

Heteroleptic iodo complexes of the 2,5-bis{*N*-(2,6-diisopropylphenyl)iminomethyl}-pyrrolyl ligand were successfully synthesized for calcium, strontium and barium. Similarly to the divalent lanthanide compounds **19-21**, all complexes **22-24** are monomeric in the solid state. Normally, strontium and barium tend to form dimeric complexes,[146, 225-229] and only few monomeric heteroleptic iodo complexes of strontium and barium are known. For example, [TpC*Sr(I)(THF)] and [TpC*Ba(I)] (TpC* = tris[3-(2-methoxy-1,1-dimethyl)-pyrazolyl]hydroborate) were synthesized with the very bulky hexadentate ligand TpC*, which saturates the coordination sphere of the corresponding metal center to form monomeric complexes.[224] In addition, the cluster compounds [IM(OtBu)$_4${Li(THF)}$_4$(OH)], described above for divalent europium, were also synthesized for calcium, strontium and barium.[207] Since the (DIP$_2$pyr)$^-$ ligand is not as bulky, it is astounding that all compounds **22-24** form monomeric complexes and the coordination sphere of each metal center is satisfied by additionally coordinated THF molecules. Remarkably, the (DIP$_2$pyr)$^-$ ligand of the calcium compound **24** exhibits an η^2-coordination mode in the solid state, which is contrary to the asymmetrically η^3-coordinated ligand in the analogous ytterbium compound **21**.

Computational studies were performed by R. Köppe in order to obtain more insight into the coordination mode of the (DIP$_2$pyr)$^-$ ligand in the calcium complex **24** and the barium derivative **23**. The calculations were performed on an *Altix 4700* (project h1191) at *"Leibniz-Rechenzentrum der Bayerischen Akademie der Wissenschaften"*. The structures of **23** and **24** were not simplified for the calculations. The possible minima in the corresponding energy hyperface were calculated within the framework of Density Functional Theory (RI-DFT)[230, 231] at the BP86-level[232-235] (basis set of def2-SVP quality for each atom[236-238]) and all of the calculations entailed the use of the TURBOMOLE package.[239] The energy differences between the symmetrically (SYM) coordinated (DIP$_2$pyr^{SYM})$^-$ ligand and the asymmetrically (ASYM) coordinated (DIP$_2$pyr^{ASYM})$^-$ ligand for both the barium and the calcium compound were calculated. In both cases, the symmetrically coordinated solvent-free species [(DIP$_2$pyr^{SYM})MI] + 4 THF exhibit a lower energy than the asymmetrically coordinated ones [(DIP$_2$pyr^{ASYM})MI] + 4 THF (both surrounded by four non coordinated THF molecules) (ΔE = 57 kJ/mol (M = Ba), 49 kJ/mol (M = Ca)). No minimum in the energy hyperface was observed for a potential asymmetrically coordinated barium compound [(DIP$_2$pyr^{ASYM})BaI(THF)$_4$]. The species [(DIP$_2$pyr^{SYM})BaI(THF)$_4$] with the structure of **23**, is calculated to be the most stable one and for a potential [(DIP$_2$pyr^{ASYM})BaI(THF)$_3$] + THF a higher energy was observed (ΔE = 31 kJ/mol). For calcium, calculations of both, the symmetrically and the asymmetrically coordinated species [(DIP$_2$pyr^{SYM})CaI(THF)$_3$] and [(DIP$_2$pyr^{ASYM})CaI(THF)$_3$] gave the same energies and no coordination mode is theoretically preferred. The asymmetrically η^2-coordinated (DIP$_2$pyr)$^-$ ligand observed in the solid state structure of **24** could be a result of temperature or/and crystal packing effects. As described earlier in this section, VT-^{1}H NMR studies showed no evidence of an asymmetrical coordination mode of the (DIP$_2$pyr)$^-$ ligand in solution even at a low temperature of -70 °C.

Moreover, experimental attempts of adding a Lewis acid, such as BPh$_3$ or AlCl$_3$ to the Schiff-base function of **24**, which is non coordinated in the solid state, failed. The fact that no reaction occurred indicate probably the coordination of both Schiff-base functions to the metal center in solution. These observations are consistent with the VT-^{1}H NMR studies and with the results of the computational calculations, which show that none of both coordination modes is energetically favored.

In order to obtain catalytically active species, the iodo ligand of **22** and **24** was replaced by the {N(SiMe$_3$)$_2$}$^-$ ligand, which has been shown to act as a leaving group in the hydroamination catalyzed by alkaline earth metal compounds.[42, 97, 99, 101] As shown in Scheme 27, the reaction of [(DIP$_2$pyr)MI(THF)$_3$] (M = Ca (**24**), Sr (**22**)) with [K{N(SiMe$_3$)$_2$}] in THF at elevated temperature afforded [(DIP$_2$pyr)M{N(SiMe$_3$)$_2$}(THF)$_2$] (M = Ca (**25**), Sr (**26**)). Since the (DIP$_2$pyr)$^-$ ligand exhibits a different coordination modes from **22** to **24**, it is remarkable that in both products **25** and **26**, the ligand is η^3-coordinated in the solid state. Differently from the analogous alkaline earth metal amides stabilized by a β-diketiminato ligand[99] (Section 1.2), no problematic ligand redistribution occurred even for the heavier homologue strontium. Only the desired heteroleptic compounds **25** and **26** were obtained.

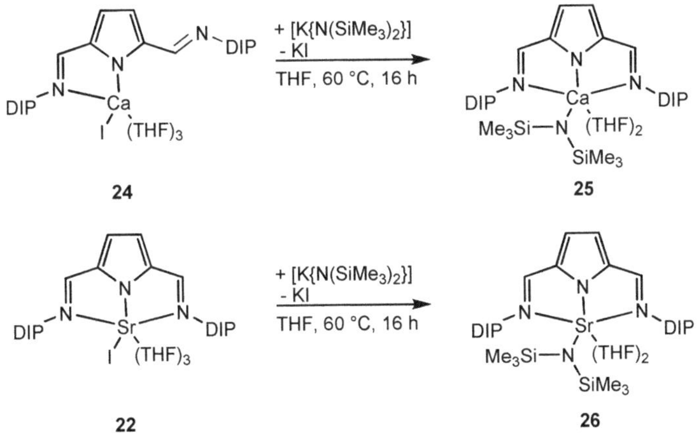

Scheme 27

Complexes **25** and **26** were characterized by standard analytical / spectroscopic techniques and the solid state structures were analyzed by single crystal X-ray diffraction. The ^{1}H NMR spectra of **25** and **26** show one singlet for the $\{N(SiMe_3)_2\}^-$ ligand (δ = -0.21 (**25**), -0.29 (**26**) ppm). While for **25** a doublet and a septet are observed for the isopropyl groups of the (DIP$_2$pyr)$^-$ ligand (δ = 1.21, 3.07 ppm), the ^{1}H NMR spectrum of **26** shows a multiplet and a septet (1.15-1.32, 3.06 ppm), which indicates a restricted rotation of the 2,6-diisopropylaniline moieties in **26**. The η^3-coordination mode is confirmed by the singlet for the imino N=CH groups (δ = 8.07 (**25**), 8.03 (**26**) ppm). The ^{13}C{^{1}H} NMR data are consistent with the information obtained from the ^{1}H NMR spectra.

Compounds **25** and **26** are isostructural in the solid state and crystallize in the monoclinic space group $P2_1/c$ with four molecules of the corresponding complex and four THF molecules in the unit cell. The solid state structure of **25** is illustrated in Figure 18 and selected bond lengths and angles are given for compounds **25** and **26**. The (DIP$_2$pyr)$^-$ ligands in **25** and **26** exhibit an η^3-coordination mode. While in **26** the ligand coordinates almost symmetrically to the metal center (Sr-N1 3.018(3) Å, Sr-N3 2.979(3) Å), the Schiff-base functions of the ligand in **25** coordinate in an asymmetrical fashion (Ca-N1 3.022(3) Å, Ca-N3 2.888(3) Å), which is certainly caused by the smaller ionic radius of Ca^{2+}. The metal center of each complex is coordinated by the (DIP$_2$pyr)$^-$ ligand, the $\{N(SiMe_3)_2\}^-$ ligand and two THF molecules, building a six-fold coordination sphere. The ligands are arranged in a distorted octahedral geometry and the apexes of each polyhedron are formed by the two THF molecules (O1-Ca-O2 171.96(11)° (**25**) and O1-Sr-O2B 174.8(2)° (**26**)). The equatorial plane of each octahedron is planar, which is confirmed by the sums of the corresponding four valence angles (360.0° (**25** and **26**)). The M-N bond distances to the $\{N(SiMe_3)_2\}^-$ ligands in **25** and **26** (Ca-N4 2.372(3) Å (**25**) and Sr-N4 2.479(3) Å (**26**)) are similar to the ones observed for the pyrrolyl units of the (DIP$_2$pyr)$^-$ ligands (Ca-N2 2.374(3) Å (**25**) and Sr-N2 2.477(3) Å (**26**)). In addition, the distorted geometry of the $\{N(SiMe_3)_2\}^-$ ligands is shown by the different M-N4-Si angles (Ca-N4-Si1 113.86(15)° *versus* Ca-N4-Si2 122.55(14)° (**25**); Sr-N4-Si1 111.83(15)° *versus* Sr-N4-Si2 121.4(2)° (**26**)). The difference between the two angles is smaller than that observed for [(C$_5$Me$_5$)$_2$Nd{CH(SiMe$_3$)$_2$}], in which agostic interactions between the neodymium ion and one methyl group of the {CH(SiMe$_3$)$_2$}$^-$ ligand were observed.[240] As a reason for the slightly asymmetrical arrangement of the $\{N(SiMe_3)_2\}^-$ ligands in **25** and **26** crystal packing effects are suggested.

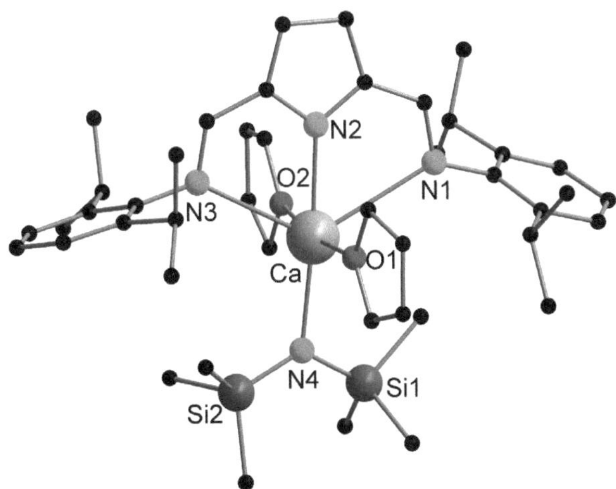

Figure 18 Solid-state structure of **25**, omitting hydrogen atoms. Selected bond lengths [Å] or angles [°] are given for **25** and the isostructural complex **26**.

25: Ca-N1 3.022(3), Ca-N2 2.374(3), Ca-N3 2.888(3), Ca-N4 2.372(3), Ca-O1 2.368(2), Ca-O2 2.363(3); N1-Ca-N2 61.08(8), N1-Ca-N3 124.98(7), N1-Ca-N4 123.60(9), N2-Ca-N3 63.90(8), N2-Ca-N4 174.96(11), N3-Ca-N4 111.42(9), N1-Ca-O1 85.23(9), N1-Ca-O2 87.42(10), N2-Ca-O1 86.85(9), N2-Ca-O2 86.75(10), N3-Ca-O1 92.98(9), N3-Ca-O2 88.55(10), N4-Ca-O1 91.62(9), N4-Ca-O2 95.17(10), O1-Ca-O2 171.96(11), Ca-N4-Si1 113.86(15), Ca-N4-Si2 122.55(14).

26 (The THF molecule around O2 is disordered and selected bond lengths [Å] and angles [°] are given for both positions O2A and O2B): Sr-N1 3.018(3), Sr-N2 2.477(3), Sr-N3 2.979(3), Sr-N4 2.479(3), Sr-O1 2.488(3), Sr-O2A 2.501(9), Sr-O2B 2.469(9), Sr-C31 3.279(5); N1-Sr-N2 60.14(9), N1-Sr-N3 121.30(8), N1-Sr-N4 125.16(10), N2-Sr-N3 61.17(9), N2-Sr-N4 174.27(10), N3-Sr-N4 113.52(9), N1-Sr-O1 86.08(10), N1-Sr-O2A 82.7(2), N1-Sr-O2B 92.6(2), N2-Sr-O1 86.30(10), N2-Sr-O2A 84.8(2), N2-Sr-O2B 88.7(2), N3-Sr-O1 91.38(11), N3-Sr-O2A 91.3(2), N3-Sr-O2B 85.1(2), N4-Sr-O1 91.80(11), N4-Sr-O2A 97.8(2), N4-Sr-O2B 93.0(2), O1-Sr-O2A 168.1(2), O1-Sr-O2B 174.8(2), Sr-N4-Si1 111.83(15), Sr-N4-Si2 121.4(2).

Catalytic applications

Since alkaline earth metal catalysts are not comprehensively studied in hydroamination reactions, the novel complexes **25** and **26** were investigated for their catalytic activity in the intramolecular hydroamination of non-activated aminoalkenes and one aminoalkynes (Table 5). Based on the observation that calcium bis(trimethylsilyl)amides are sufficiently basic to effect deprotonation of terminal alkynes to form acetylide complexes,[241] an internal alkyne was chosen for the presented studies. All experiments were carried out under rigorously anaerobic reaction conditions by using dry, degassed substrates and a catalyst loading of 5 mol % and were monitored by ^{1}H NMR spectroscopy, by using ferrocene as an internal standard. Both complexes **25** and **26** are catalytically active for the hydroamination. As observed for the aminotroponiminate alkaline earth metal amides,[42] the calcium species **25** is more active than the strontium compound **26**. This result conflicts with the observed reactivity of the bis(imidazolin-2-ylidene-1-yl)borate alkaline earth metal amides, in which the strontium derivative showed the highest activity (Section 1.2).[100] As shown in Table 5, substrates **27a**, **28a** and **29a** were cyclized selectively to the corresponding pyrrolidines **27b**, **28b** and **29b** (entries 1-6) in high yields. The aminoalkenes **27a** and **28a** which contain bulky substituents in the β-position to the amino group (Thorpe-Ingold-effect)[61-64] were converted rapidly to their cyclic products at room temperature. By using the more active catalyst **25** at room temperature, a reaction time of 8 hours was required to cyclize substrate **29a** which contains less bulky substituents (Table 5, entry 5). While by using **26** as a catalyst for the cyclization of **29a**, an elevated temperature of 60 °C is necessary and the conversion is completed after two hours (Table 5, entry 6). The best result was obtained by using **27a** as the substrate and **25** as the catalyst. The corresponding product **27b** was formed at room temperature within 2 minutes in 99 % yield (Table 5, entry 1). This result is better than that observed by using the aminotroponiminate calcium amide as a catalyst,[42] and similar to that obtained for the β-diketiminato calcium amide [{(DIPNC(Me))$_2$CH}Ca{N(SiMe$_3$)$_2$}(THF)], which cyclized **27a** within 15 minutes, but by using a catalyst loading of 2 mol % only.[99] On the contrary, the reaction of the α-branched aminoalkene **30a** and of the aminoalkyne **31a** catalyzed by either **25** or **26** did not lead selectively to the desired products **30b** and **31b** (Table 5, entries 7-11). The cyclizations were accompanied by side reactions (Scheme 28 and Scheme 29) and the products were determined *via* ^{1}H NMR spectroscopy. In the reaction of the α-branched aminoalkene **30a**, competitive alkene isomerization occurs in all reactions (Table 5, entries 7-9). As shown in Scheme 28, the reaction of **30a** with either catalyst **25** or **26** afforded the desired product **30b** and additionally 2-amino-4-hexene (**30c**). Since internal alkenes such as **30c** are less reactive than terminal aminoalkenes, **30c** is not cyclized by the catalysts. The alkene isomerization depends on the reaction temperature and the type of catalyst. While in the reaction of **30a** catalyzed by **25** within 46 hours at room temperature 18 % of the undesired product **30c** was observed (Scheme 28, entry 7), at an elevated temperature of 60 °C the conversion was completed after 5 hours and 40 % of **30c** was detected (Scheme 28, entry 8). In addition, the elevated temperature decreases the *cis/trans* selectivity from *cis*:*trans* = 1:13 at room temperature to 1:8 at 60 °C. By using **26** as a catalyst, the alkene isomerization reaction is favored and even at room temperature a yield of

71 % of the undesired product **30c** was determined (Scheme 28, entry 9). As described in Section 1.2, a similar alkene isomerization was observed by Hill *et al.* by using [{(DIPNC(Me))$_2$CH}Ca{N(SiMe$_3$)$_2$}(THF)] as a catalyst and the proposed reaction mechanism is shown in Section 1.2, Scheme 4.

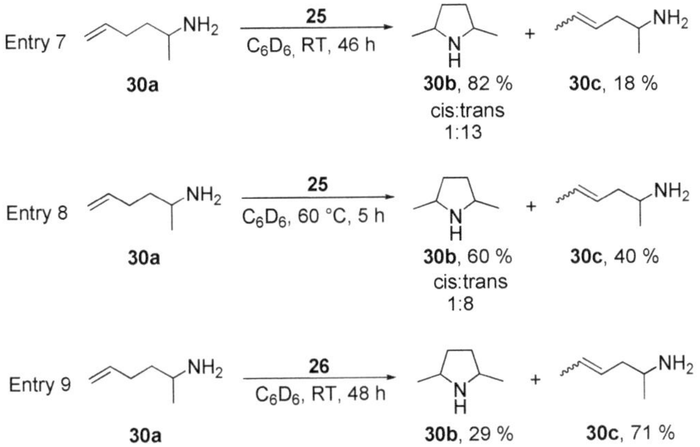

Entry 7

30a

25 / C$_6$D$_6$, RT, 46 h

30b, 82 %
cis:trans
1:13

+

30c, 18 %

Entry 8

30a

25 / C$_6$D$_6$, 60 °C, 5 h

30b, 60 %
cis:trans
1:8

+

30c, 40 %

Entry 9

30a

26 / C$_6$D$_6$, RT, 48 h

30b, 29 %

+

30c, 71 %

Scheme 28

By using the aminoalkyne **31a** as a substrate, both products derived from the imine-enamine tautomeric equilibrium were obtained. In addition to the desired product **31b**, the cyclic enamine **31c** was detected (Scheme 29). Independently from the catalyst used for the reaction of **31a**, 20 % of the byproduct **31c** was found. The conversion was completed after 1 hour at room temperature by using **25** as a catalyst (Scheme 29, entry 10); while for the reaction of **31a** catalyzed by **26**, a reaction time of 5 hours was required (Scheme 29, entry 11).

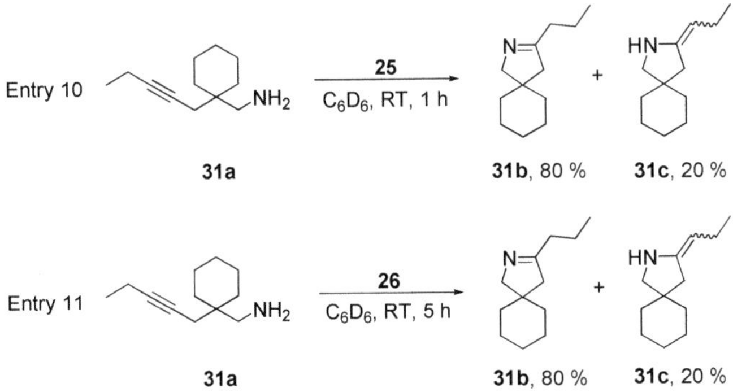

Entry 10

31a

25 / C$_6$D$_6$, RT, 1 h

31b, 80 %

+

31c, 20 %

Entry 11

31a

26 / C$_6$D$_6$, RT, 5 h

31b, 80 %

+

31c, 20 %

Scheme 29

Since aminoalkenes are less reactive than aminoalkynes, **31a** was expected to undergo the cyclization more quickly than the related aminoalkene **28a**. The slower reaction of **31a** (Table 5, entries 9 and 10) compared with that of **28a** (Table 5, entries 3 and 4) could be a result of the side reaction occurred in the cyclization of **31a**, which could influence the reaction time. The reason for the shifted imine-enamine tautomeric equilibrium is unclear. In general the original species obtained from the catalytic cycle is an enamine. If this enamine is not tertiary, it will be tautomerized into the corresponding imine.[70] The state of the tautomeric equilibrium usually lies on the side of the imine. In the published hydroamination of substrate **31a** catalyzed by rare earth metal amides, only the imine **31b** was formed.[156] In addition, **31a** was used for further hydroamination studies presented in this work (Section 3.3) and no formation of an enamine was observed.

The reaction scope of the hydroamination catalyzed by **25** and **26** is limited, since undesired side reactions occurred for the substrates **30a** and **31a**. Nevertheless, both complexes **25** and **26** were proven to be catalytically active systems for the intramolecular hydroamination of aminoalkenes. In particular, the calcium complex **25** showed high activity and can be compared with the β-diketiminato calcium amide $[\{(DIPNC(Me))_2CH\}Ca\{N(SiMe_3)_2\}(THF)]$, the aminotroponiminate calcium amide $[\{(iPr)_2ATI\}Ca\{N(SiMe_3)_2\}(THF)_2]$ $((iPr)_2ATI = N$-isopropyl-2-(isopropylamino)tropon-iminate) and the bis(imidazolin-2-ylidene-1-yl)borate alkaline earth metal amides $[\{H_2B(Im^{tBu})_2\}M\{N(SiMe_3)_2\}(THF)_n]$ (M = Ca, n = 1; M = Sr, n = 2; Im^{tBu} = 1-*tert*-butylimidazole),.[42, 99, 100] In order to obtain more information about the activity of the calcium species **25**, further hydroamination studies are required for example by using aminoalkenes with an internal C-C double bond which normally undergo the cyclization slower than aminoalkenes with a terminal double bond. Furthermore, substrates which form six or seven membered rings by cyclization are more difficult to react. To expand the investigations of **25** to such substrates would be a great challenge. In addition, Hill *et al.* observed an increased reactivity by increasing the concentration of the catalyst.[99] Therefore, another prospective task will be to study the reactivity of **25** by using different catalyst concentrations.

Table 5 Hydroamination reactions of terminal aminoalkenes and alkynes catalyzed by **25** and **26**.[a]

Entry	Substrate	Product	Cat (5 mol %)	T [°C]	t [h]	Yield[b] [%]
1	**27a**	**27b**	25	r.t.	0.03	99
2			26	r.t.	0.5	98
3	**28a**	**28b**	25	r.t.	0.33	quant
4			26	r.t.	4	quant
5	**29a**	**29b**	25	r.t	8	96
6			26	60	2	92
7	**30a**	**30b**	25	r.t.	46	82[c] (1:13)[d]
8			25	60	5	60[c] (1:8)[d]
9			26	r.t.	48	29[c] (-)[e]
10	**31a**	**31b**	25	r.t.	1	80[f]
11			26	r.t.	5	80[f]

[a]Conditions: Cat. 15-20 mg (5 mol %), C_6D_6; [b]calculated by ^{1}H NMR, ferrocene as internal standard; [c]products were accompanied by alkene isomerization byproduct **30c** (see text for details); [d]ratio *cis:trans*; [e]*cis/trans*-signals obscured by other signals; [f]products were accompanied by the enamine **31c** (see text for details).

Derivatives of the divalent lanthanides analogous to **25** and **26** were hitherto not synthesized. Since Marks *et al.* showed that divalent lanthanide complexes form trivalent species by oxidation during the catalytic cycle,[33] the synthesis of potential [(DIP$_2$pyr)Ln{N(SiMe$_3$)$_2$}(THF)$_n$] as catalysts for hydroamination reactions was not of great interest. However, it would be interesting to know whether the heteroleptic species can be formed or if the homoleptic complex [(DIP$_2$pyr)$_2$Ln(THF)$_n$] would be preferred by the reaction of [(DIP$_2$pyr)LnI(THF)$_3$] with [K{N(SiMe$_3$)$_2$}]. As observed earlier by our group, the reactions of lanthanide bis(phosphinimino)methanide iodo complexes with [K{N(SiMe$_3$)$_2$}] afforded the homoleptic complexes [Ln{CH(PPh$_2$NSiMe$_3$)$_2$}$_2$] and [Ln{N(SiMe$_3$)$_2$}$_2$(THF)$_n$] (Ln = Sm, Yb).[43, 148, 242] On the contrary, by using the aminotroponiminate ligand, the heteroleptic amide complexes [{(*i*Pr)$_2$ATI}Ln{N(SiMe$_3$)$_2$}(THF)$_2$] (Ln = Eu, Yb) were formed in a convenient one pot reaction of [{(*i*Pr)$_2$ATI}K], anhydrous lanthanide diiodides and [K{N(SiMe$_3$)$_2$}].[42] Since the divalent lanthanides and the alkaline earth metals tend to form similar complexes, the formation of heteroleptic complexes [(DIP$_2$pyr)Ln{N(SiMe$_3$)$_2$}(THF)$_n$] is expected in the present case. In addition, it might be attractive to obtain the homoleptic complexes [(DIP$_2$pyr)$_2$Ln(THF)$_n$], which are hitherto unknown for divalent lanthanides, since they could exhibit interesting coordination modes of the ligand.

In summary, the 2,5-bis{N-(2,6-diisopropylphenyl)iminomethyl}pyrrolyl ligand was shown to stabilize complexes of the divalent lanthanides and alkaline earth metals. Heteroleptic iodo complexes of divalent samarium, europium, ytterbium, calcium, strontium and barium were successfully synthesized. Surprisingly, all complexes are monomeric in the solid state, independently from the ionic radii of their center metals. The coordination sphere of each complex is satisfied by additionally coordinated THF molecules. This structural motif is very rare in the chemistry of the larger divalent lanthanides and alkaline earth metals and, to the best of my knowledge, it was observed (except for cluster compounds or complexes with very bulky ligands)[207, 208, 224] only one time previously.[217] In addition, the 2,5-bis{N-(2,6-diisopropylphenyl)iminomethyl}pyrrolyl ligand showed interesting coordination modes. Except for the calcium iodo complex, an η^3-coordination mode was observed for the (DIP$_2$pyr)$^-$ ligand in all compounds presented in this work. While the (DIP$_2$pyr)$^-$ ligand of the ytterbium iodo complex coordinates in an asymmetrical η^3-fashion, the analogous calcium derivative is η^2-coordinated in the solid state. Computational studies of the calcium species showed the same energy for both, a symmetrically η^3- and an asymmetrically η^2-coordinated ligand and thus the preferred η^2-coordination mode observed in the solid state structure of the calcium complex is probably a result of temperature or/and crystal packing effects. In addition, VT-^{1}H NMR studies of the calcium iodo complex showed no evidence of an asymmetrically coordinated ligand in solution. Furthermore, catalytically active calcium and strontium species were prepared by introducing the {N(SiMe$_3$)$_2$}$^-$ ligand as a leaving group and the resulting compounds were investigated for the intramolecular hydroamination of aminoalkenes and one aminoalkyne. The formation of undesired side products by alkene isomerization and imine-enamine tautomerism were observed in two cases for both catalysts, which led to a limited reaction scope. However, both derivatives were proven to be catalytically active and the best results were obtained for the calcium complex, which shows an activity comparable to that of the bis(imidazolin-2-ylidene-1-yl)borate alkaline earth metal amides, the β-diketiminato calcium amide [{(DIPNC(Me))$_2$CH}Ca{N(SiMe$_3$)$_2$}(THF)] and the aminotroponiminate calcium amide [{(iPr)$_2$ATI}Ca{N(SiMe$_3$)$_2$}(THF)$_2$].[42, 99, 100]

3.3   Hydroamination studies of rare earth metal complexes coordinated by imidazolin-2-imide derivatives

Imidazolin-2-imides were shown to be versatile ligands in transition metal chemistry.[243-246] They are synthesized by deprotonation of the corresponding imidazolin-2-imines which can be easily obtained by a Staudinger-like reaction between N-heterocyclic carbenes[247, 248] and trimethylsilyl azide followed by desilylation of the N-silylated 2-iminoimidazoline intermediates in methanol.[246, 249] The ability of the imidazolium ring to stabilize a positive charge leads to highly basic ligands with a strong electron-donating capacity towards transition metals and thus imidazolin-2-imides can be regarded as monodentate analogues of cyclopentadienyl ligands. Imidazolin-2-imido transition metal complexes are suitable catalysts for reactions like the polymerization of ethylene or alkyne cross-metathesis reactions.[250-253] In addition to serve as ligand precursors for the corresponding imidazolin-2-imides, imidazolin-2-imines can also be used to synthesize novel multidentate poly(imidazolin-2-imine) ligands or cyclopentadienyl-imidazolin-2-imines.[243, 254-256] M. Tamm et $al.$ introduced the 1,3-bis(2,6-diisopropylphenyl)-imidazolin-2-imide {Im^{DIP}N}$^-$ as well as a neutral bis(imidazolin-2-imino)pyridine pincer ligand into the rare earth metal chemistry.[257, 258] In the presented work, novel rare earth metal alkyl complexes coordinated by either a cyclopentadienyl-imidazolin-2-imine or the {Im^{DIP}N}$^-$ ligand were synthesized by M. Tamm et $al.$ and should now be investigated for the intramolecular hydroamination of aminoalkenes and aminoalkynes.[259, 260]

The synthesis of the rare earth metal alkyl complexes $[(\eta^5:\eta^1\text{-}C_5Me_4\text{-}SiMe_2\text{-}NIm^{iPr})Ln(CH_2SiMe_3)_2]$ (Ln = Sc (**35a**), Y (**35b**), Lu (**35c**)) (Im^{iPr} = 1,3-diisopropyl-4,5-dimethylimidazolin) was performed by M. Tamm *et al.* (Scheme 30).[259] The reaction of 1,3-diisopropyl-4,5-dimethylimidazolin-2-imine (**32**) with **33** afforded the neutral imidazolin-2-imino-tetramethylcyclopentadiene ligand **34**. By deprotonation of **34** with $[Ln(CH_2SiMe_3)_3(THF)_2]$ (Ln = Sc, Y, Lu), the products **35a-35c** were obtained *via* alkyl elimination (Scheme 30).

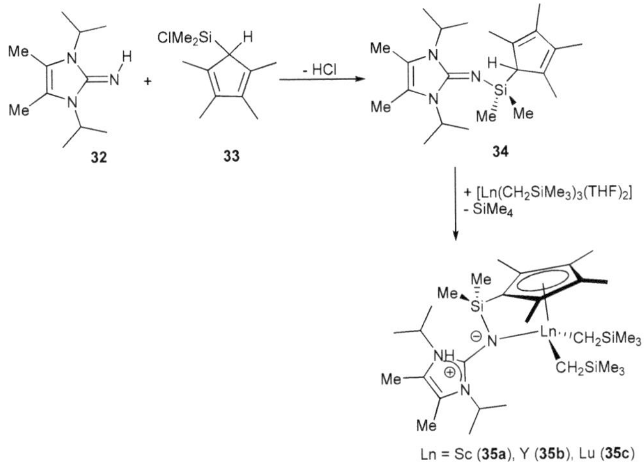

Scheme 30

In order to evaluate the catalytic activity of compounds **35a–35c** in comparison to other rare earth metal complexes, they were investigated for the hydroamination of non-activated aminoalkenes and one aminoalkyne.[259] Compounds **35a–35c** showed high activities and all substrates were converted regiospecifically into the cyclic products under mild conditions (Table 6). All experiments were carried out under rigorously anaerobic reaction conditions by using dry, degassed substrates with low catalyst loadings of 2-6 mol %. Kinetic studies were undertaken by monitoring the reactions with *in situ* ^{1}H NMR spectroscopy until substrate consumptions were complete and the decrease of the substrate peaks was integrated *versus* the product signals. The reaction rate increases with increasing ionic radii of the center metals and the yttrium compound **35b** showed the highest activity. This is consistent with the results obtained for metallocene catalysts of the lanthanides.[33] By using **35b**, all substrates were converted at room temperature (Table 6, entries 2, 5, 9 and 14), whereas for reactions catalyzed by the scandium compound **35a** an elevated temperature of 60 °C was required (Table 6, entries 1, 4, 7 and 13). It is well known that the hydroamination of aminoalkynes proceeds significantly faster than the cyclization of aminoalkenes.[70] Consequently, the aminoalkyne **31a** undergoes the hydroamination within a few minutes (Table 6, entries 13-15). In addition, the aminoalkenes **27a** and **28a** which contain bulky substituents in the β-position to the amino group (Thorpe-Ingold-effect) were converted rapidly to their cyclic products (Table 6, entries 1-3 and 4-6). The α-branched aminoalkene **30a** exhibits no bulky substituents in the β-position and thus significantly lower rates were observed for all catalysts (Table 6, entries 7-12). The cyclization of substrate **30a** catalyzed by **35b** compared with the compounds $[(\eta^5{:}\eta^1\text{-}C_5Me_4\text{-}SiMe_2\text{-}N\mathit{t}Bu)LnE(SiMe_3)_2]$[261] (E = N, CH) shows a lower turnover frequency for **35b** by a factor of about three. In contrast, a comparison of **35b** with actinide catalysts in the cyclization of substrate **27a** shows a similar turnover frequency to $[(\eta^5{:}\eta^1\text{-}C_5Me_4\text{-}SiMe_2\text{-}N\mathit{t}Bu)U(NMe_2)_2]$ but a lower turnover frequency than $[(\eta^5{:}\eta^1\text{-}C_5Me_4\text{-}SiMe_2N\mathit{t}Bu)\text{-}Th(NMe_2)_2]$.[262]

Table 6 Hydroamination reactions of terminal aminoalkenes and alkynes catalyzed by **35a–35c**.[a]

Entry	Substrate	Product	Cat	Cat [mol%]	T [°C]	t [min]	Yield[b] [%]
1			**35a**	6	60	60	99
2	**27a**	**27b**	**35b**	4	r.t.	3	quant
3			**35c**	5	r.t.	19	99
4			**35a**	5	60	90	99
5	**28a**	**28b**	**35b**	5	r.t.	3	quant
6			**35c**	5	r.t.	17	99
7			**35a**	2	60	6000	99 (1:12)[c]
8			**35a**	2	120	3840	99 (1:12)[c]
9	**30a**	**30b**	**35b**	3	r.t.	240	99 (1:9)[c]
10			**35b**	3	60	60	99 (1:9)[c]
11			**35c**	2	r.t.	1200	99 (1:15)[c]
12			**35c**	2	60	300	99 (1:15)[c]
13			**35a**	4	60	10	quant
14	**31a**	**31b**	**35b**	5	r.t.	5	quant
15			**35c**	4	r.t.	19	quant

[a]Conditions: Cat. 10 mg, C_6D_6; [b]calculated by ^{1}H NMR; [c]ratio *cis:trans*.

Furthermore, the rare earth metal alkyl complexes [{Im^{DIP}N}Ln(CH$_2$SiMe$_3$)$_2$(THF)$_2$] (Ln = Sc (**38a**), Y (**38b**), Lu (**38c**)) were synthesized by M. Tamm *et al.* (Scheme 31).[260] The reaction of 1,3-bis(2,6-diisopropylphenyl)-imidazolin-2-imine (**36**) with [LiCH$_2$SiMe$_3$] and anhydrous lanthanide trichlorides afforded [{Im^{DIP}N}LnCl$_2$(THF)$_3$] (Ln = Sc (**37a**), Y (**37b**), Lu (**37c**)).[257] As shown in Scheme 31, complexes **37a-37c** were reacted with two equivalents of [LiCH$_2$SiMe$_3$] to give the desired compounds **38a-38c**.

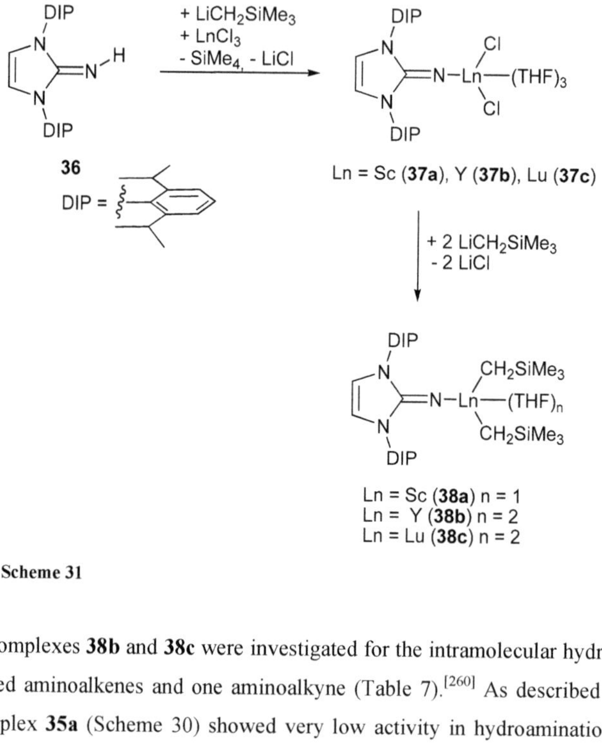

Scheme 31

The novel complexes **38b** and **38c** were investigated for the intramolecular hydroamination of non-activated aminoalkenes and one aminoalkyne (Table 7).[260] As described above, the scandium complex **35a** (Scheme 30) showed very low activity in hydroamination reactions (Table 6, entries 1, 4, 7 and 13). For this reason only the yttrium and the lutetium compounds **38b** and **38c** were studied (Table 7). All experiments were carried out under rigorously anaerobic reaction conditions by using dry, degassed substrates and a catalyst loading of 5 mol % and were monitored by ^{1}H NMR spectroscopy. The decrease of the substrate peaks was integrated *versus* the product signals and ferrocene was additionally used as an internal standard.

Table 7 Hydroamination reactions of terminal aminoalkenes and alkynes catalyzed by **38b**, **38c**.[a]

Entry	Substrate	Product	Cat (5 mol %)	T [°C]	t [min]	Yield[b] [%]
1	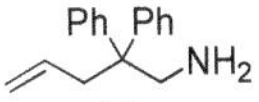 **27a**	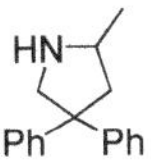 **27b**	**38b**	r.t.	5	98
2			**38c**	r.t.	5	98[d]
3	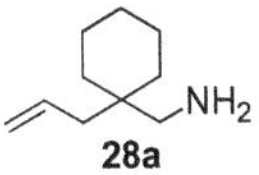 **28a**	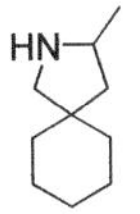 **28b**	**38b**	r.t.	5	quant[d]
4			**38c**	r.t.	5	quant
5	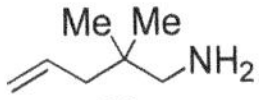 **29a**	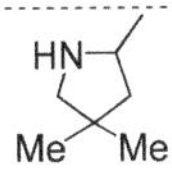 **29b**	**38b**	60	60	92[d]
6			**38c**	60	60	90
7	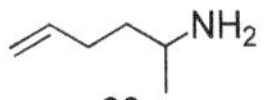 **30a**	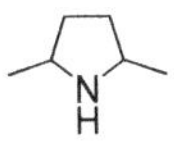 **30b**	**38b**	120	180	quant (1:3)[c]
8			**38c**	120	60	quant (1:3)[c]
9	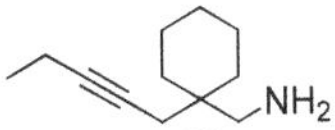 **31a**	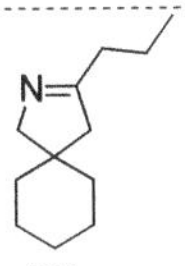 **31b**	**38b**	r.t.	5	quant
10			**38c**	r.t.	5	quant

[a]Conditions: Cat. 15-20 mg (5 mol %), C_6D_6; [b]calculated by ^{1}H NMR; [c]ratio *cis:trans*; [d]ferrocene as internal standard.

Both complexes **38b** and **38c** were shown to be active catalysts and all substrates were cyclized regiospecifically. Differently from **35a-35c**, no trend of the reaction rate dependent on the ionic radii of the center metals was observed. The activity of **38b** and **38c** is very similar, except for the reaction of the α-branched aminoalkene **30a**, which is cyclized faster by the lutetium derivative **38c** than by the yttrium compound **38b** (Table 7, entries 7 and 8). The aminoalkyne **31a** was converted quantitatively within 5 minutes at room temperature to the cyclic product (Table 7, entries 9 and 10). Furthermore, the aminoalkenes **27a** and **28a** which contain bulky substituents in the β-position to the amino group were also cyclized within 5 minutes at room temperature in 98 % to quantitative yields (Table 7, entries 1-4). The aminoalkene **29a** exhibits less bulky substituents in the β-position to the amino group and an elevated reaction temperature of 60 °C is required. Both catalysts **38b** and **38c** were shown to cyclize **29a** within 60 minutes and 90-92 % of the cyclic product was observed (Table 7, entries 5 and 6). Since the α-branched aminoalkene **30a** contains no bulky substituents in the β-position to the amino group, the reactions show significantly lower rates and require a temperature of 120 °C and longer reaction times as well (Table 7, entries 7 and 8). The yttrium catalyst **38b** shows similar activities to **35b** (Table 6). A higher activity for **35b** is observed only if the less reactive substrate **30a** was used (Table 6, entries 9 and 10 *versus* Table 7, entry 7). On the contrary, the lutetium catalyst **38c** is superior to **35c** and the cyclizations proceed faster. The rates for the intramolecular hydroamination reactions catalyzed by **38b** and **38c** were lower than those observed for the corresponding metallocene catalysts, for example [(C$_5$Me$_5$)$_2$Ln{CH(SiMe$_3$)$_2$}] (Ln = La, Sm, Nd).[70] However, complex **38b** showed significantly higher activity for the cyclization of substrates **27a**, **28a** and **31a** compared with the non-cyclopentadienyl complex [{(Me$_3$SiNPPh$_2$)$_2$CH}La-{N(SiHMe$_2$)$_2$}$_2$].[156]

In summary, both the imidazolin-2-imide and the cyclopentadienyl-imidazolin-2-imine ligands were proven to stabilize rare earth metal complexes and catalytically active species were formed. The corresponding alkyl derivatives showed high selectivities and activities in the intramolecular hydroamination of non-activated aminoalkenes and aminoalkynes. The novel compounds cannot compete with the metallocene analogues, but in particular the imidazolin-2-imide complexes are new examples for post-metallocenes, serving as highly active catalysts in hydroamination reactions.

4.   Experimental section

4.1   General considerations

All manipulations of air-sensitive materials were performed with the rigorous exclusion of oxygen and moisture in flame-dried Schlenk-type glassware either on a dual manifold Schlenk line, interfaced to a high vacuum (10^{-3} torr) line, or in an argon-filled MBraun glove box. NMR spectra were recorded on Jeol JNM-LA 400 FT-NMR, a Bruker Avance 400 MHz or a Bruker Avance II 300 MHz NMR spectrometer. Chemical shifts are referenced to internal solvent resonances and are reported relative to tetramethylsilane, 85 % phosphoric acid (^{31}P NMR) and 15% $BF_3 \cdot Et_2O$ (^{11}B NMR), respectively. IR spectra were obtained on a FTIR Spektrometer Bruker IFS 113v. Elemental analyses were carried out with an Elementar vario EL or EL III. Ether solvents (THF and Et_2O) were predried over Na wire or by using an MBraun solvent purification system (SPS-800) and distilled under nitrogen from potassium benzophenone ketyl prior to use. Hydrocarbon solvents (toluene and *n*-pentane) were distilled under nitrogen from $LiAlH_4$ or were dried by using an MBraun solvent purification system (SPS-800). All solvents for vacuum line manipulations were stored *in vacuo* over $LiAlH_4$ in resealable flasks. Deuterated solvents were obtained from Aldrich (99 atom % D) and were degassed, dried and stored *in vacuo* over Na/K alloy in resealable flasks. $LnCl_3$,[263] LnI_2,[264-266] $[Ln(BH_4)_3(THF)_3]$,[108] $CH_2(PPh_2NSiMe_3)_2$,[267] $[K\{CH(PPh_2NSiMe_3)_2\}]$ (1),[153] $[\{(Me_3SiNPPh_2)_2CH\}YCl_2]$ (2),[154] $(DIP_2pyr)H$ (8),[268, 269] $[(DIP_2pyr)K]$ (10),[163, 164] $[(DIP_2pyr)Ln(BH_4)_2(THF)_2]$ (Ln = La (13), Nd (14))[164, 200] and $[\{DIP_2pyr^*-BH_3\}Lu(BH_4)(THF)_2]$ (16)[164, 200] were prepared according to literature procedures.

4.2 Synthesis of the new compounds

4.2.1 Synthesis of bis(phosphinimino)methanide rare earth metal borohydrides

[{(Me$_3$SiNPPh$_2$)$_2$CH}La(BH$_4$)$_2$(THF)] (**3**)

THF (25 ml) was condensed at -78 °C onto a mixture of [La(BH$_4$)$_3$(THF)$_3$] (600 mg, 1.50 mmol) and [K{CH(PPh$_2$NSiMe$_3$)$_2$}] (**1**) (895 mg, 1.50 mmol) and the resulting reaction mixture was stirred for 16 h at 60 °C. The colorless solution was filtered and the solvent evaporated *in vacuo*. Then, toluene (10 ml) was condensed onto the residue. The mixture was heated carefully until the solution became clear. The solution was allowed to stand at room temperature to obtain the product as colorless powder after 16 h. - Yield: 842 mg, 1.10 mmol, 74 %. Single crystals were obtained by crystallization from hot THF. - ^{1}H NMR (THF-d$_8$, 400 MHz, 25 °C): δ = 0.22 (s, 18 H, SiMe$_3$), 1.02-1.64 (br, 8 H, BH$_4$), 2.02 (t, 1 H, CH, $J_{H,P}$ = 2.1 Hz), 7.41-7.84 (m, 20 H, Ph) ppm. - ^{13}C{^{1}H} NMR (THF-d$_8$, 100.4 MHz, 25 °C): δ = 4.3 (SiMe$_3$), 16.6 (CH), 128.2-128.4 (m, m-Ph), 130.8 (p-Ph), 131.0-131.1 (m, o-Ph), 131.9-132.2 (m, i-Ph) ppm. - ^{31}P{^{1}H} NMR (THF-d$_8$, 101.3 MHz, 25 °C): δ = 15.6 ppm. - ^{11}B NMR (THF-d$_8$, 128.15 MHz, 25 °C): δ = - 22.4 (br qt, $J_{B,H}$ = 83 Hz) ppm. - IR (ν/cm^{-1}): 693 (s), 832 (vs), 931 (m), 1072 (s), 1094 (s), 1131 (s), 1161 (w), 1254 (m), 1434 (m), 1976 (w), 2151 (m), 2210 (m), 2424 (w), 2949 (w), 3056 (w). - EI/MS (70 eV, 180 °C) *m/z* (%): 712 (M$^+$ - THF - BH$_4$, 33), 698 ({(Me$_3$SiNPPh$_2$)$_2$CH}La$^+$, 14), 581 (25), 569 (97), 569 (97), 558 (CH(PPh$_2$NSiMe$_3$)$_2^+$, 84), 544 (CH(PPh$_2$NSiMe$_3$)$_2^+$ - CH$_3$, 100), 493 (85), 481 (23), 455 (59) 394 (36), 348 ([M$^+$ - THF - 2 BH$_4$] / 2, 11), 287 (26), 272 (Ph$_2$PNSiMe$_3$, 48), 183 (43), 135 (73), 121 (90), 73 (SiMe$_3$, 60), 43 (20). - (**3** - 1/2 THF) C$_{33}$H$_{51}$B$_2$N$_2$O$_{1/2}$Si$_2$P$_2$La (762.73): calcd. C, 51.99, H, 6.74, N, 3.67; found C, 51.43, H, 7.32, N, 2.72 %.

[{(Me$_3$SiNPPh$_2$)$_2$CH}Nd(BH$_4$)$_2$(THF)] (**4**)

THF (25 ml) was condensed at -78 °C onto a mixture of [Nd(BH$_4$)$_3$(THF)$_3$] (740 mg, 1.83 mmol) and [K{CH(PPh$_2$NSiMe$_3$)$_2$}] (**1**) (1.09 g, 1.50 mmol) and the resulting blue reaction mixture was stirred for 16 h at 60 °C. The blue solution was filtered off, concentrated and layered with *n*-pentane. The product was obtained as purple crystals after 16 h at room temperature. - Yield: 704 mg, 0.80 mmol, 44 % (single crystals). - IR (ν/cm^{-1}): 708 (s), 750 (w), 833 (s), 930 (m), 1016 (w), 1070 (s), 1093 (s), 1125 (s), 1159 (w), 1248 (m), 1433 (w), 1729 (w), 1995 (w), 2078 (m), 2121 (m), 2204 (w), 2334 (m), 2418 (w), 2924 (w). - EI/MS (70 eV, 180 °C) *m/z* (%): 717 (M$^+$ - THF - BH$_4$, 9), 702 ({(Me$_3$SiNPPh$_2$)$_2$CH}Nd$^+$, 2),

569 (81), 558 ($CH(PPh_2NSiMe_3)_2^+$, 94), 543 ($CH(PPh_2NSiMe_3)_2^+$ - CH_3, 100), 493 (44), 481 (21), 455 (58) 394 (34), 345 ([M^+ - CH_3 - 2 BH_4] / 2, 29), 287 (26), 272 ($Ph_2PNSiMe_3$, 40), 264 (25), 256 (19), 199 (26), 183 (30), 135 (53), 121 (82), 73 ($SiMe_3$, 38). - (**4** + 1 THF) $C_{39}H_{63}B_2N_2O_2Si_2P_2Nd$ (875.91): calcd. C, 53.48, H, 7.25, N, 3.20; found C, 53.87, H, 7.10, N, 2.92 %.

[{($Me_3SiNPPh_2)_2CH$}Sc($BH_4)_2$] (**5**)

THF (25 ml) was condensed at -78 °C onto a mixture of [Sc($BH_4)_3(THF)_2$] (97 mg, 0.41 mmol) and [K{$CH(PPh_2NSiMe_3)_2$}] (**1**) (191 mg, 0.32 mmol) and the resulting reaction mixture was stirred for 16 h at 60 °C. The colorless solution was filtered and the solvent evaporated *in vacuo*. Then, toluene (20 ml) was condensed onto the residue. The solution was filtered again and concentrated until a colorless precipitate appears. The mixture was heated carefully until the solution became clear. The solution was allowed to stand at room temperature to obtain the product as colorless crystals after 16 h. - Yield: 129 mg, 0.18 mmol, 56 % (single crystals). - ^1H NMR (THF-d_8, 400 MHz, 25 °C): δ = 1.09 (s, 18 H, $SiMe_3$), 0.99-1.41 (br, 8 H, BH_4), 2.06 (t, 1 H, CH, $J_{H,P}$ = 3.4 Hz), 2.42 (s, toluene), 7.19-7.24 (m, 4 H, Ph), 7.28 - 7.31 (m, toluene), 7.39-7.43 (m, 6 H, Ph), 7.57-7.67 (m, 6 H, Ph), 7.98-8.03 (m, 4 H, Ph) ppm. - $^{13}C\{^1H\}$ NMR (THF-d_8, 100.4 MHz, 25 °C): δ = 3.1 ($SiMe_3$), 16.3 (t, CH, $J_{C,P}$ = 98 Hz), 21.2 (toluene), 125.7 (toluene), 128.6 (toluene), 128.7, 128.9 (2 t, m-Ph, $J_{C,P}$ = 6.0 Hz), 129.3 (toluene), 131.7 (p-Ph), 132.0 (t, o-Ph, $J_{C,P}$ = 5.2 Hz), 132.4 (p-Ph), 132.7 (t, o-Ph, $J_{C,P}$ = 5.2 Hz), 134.4 (m, i-Ph), 135.4 (m, i-Ph), 138.1 (toluene) ppm. - $^{31}P\{^1H\}$ NMR (THF-d_8, 101.3 MHz, 25 °C): δ = 19.9 ppm. - ^{11}B NMR (THF-d_8, 128.15 MHz, 25 °C): δ = -17.5, -22.4 ppm. - IR (v/cm^{-1}): 715 (s), 740 (m), 835 (vs), 950 (m), 1057 (m), 1106 (m), 1157 (w), 1246 (w), 1436 (m), 1589 (w), 1870 (w), 1991 (w), 2078 (m), 2291 (w), 2352 (w), 2495 (w), 2922 (w). - EI/MS (70 eV, 180 °C) m/z (%): 675 (13), 638 (12), 632 (M^+, 1), 618 (M^+ - BH_4, 67), 603 ({($Me_3SiNPPh_2)_2CH$}Sc$^+$, 17), 569 (42), 562 (12), 558 ($CH(PPh_2NSiMe_3)_2^+$, 14), 543 ($CH(PPh_2NSiMe_3)_2^+$ - CH_3, 100), 493 (22), 455 (13), 394 (6), 301 ([M^+ - 2 BH_4] / 2, 7), 272 ($Ph_2PNSiMe_3$, 26), 197 (10), 135 (17), 121 (14), 73 ($SiMe_3$, 25), 43 (13). - (**5** + 1/2 toluene) $C_{34.5}H_{51}B_2N_2Si_2P_2Sc$ (678.49): calcd. C, 61.07, H, 7.58, N, 4.13; found C, 60.84, H, 7.57, N, 3.90 %.

$[\{(Me_3SiNPPh_2)_2CH\}Y(BH_4)_2]$ (**6**)

THF (25 ml) was condensed at -78 °C onto a mixture of $NaBH_4$ (42 mg, 1.10 mmol) and *in situ* prepared $[\{(Me_3SiNPPh_2)_2CH\}YCl_2]_2$ (**2**) (0.25 mmol) and the resulting colorless reaction mixture was stirred for 16 h at 60 °C. The colorless solution was filtered off and concentrated until a precipitate appears. The mixture was heated carefully until the solution became clear. The solution was allowed to stand at room temperature to obtain the product as colorless crystals after 6 h. - Yield: 171 mg, 0.25 mmol, 51 % (single crystals). - ^{1}H NMR (THF-d_8, 400 MHz, 25 °C): δ = 0.19 (s, 18 H, $SiMe_3$), 1.70-2.10 (br, 8 H, BH_4), 1.94 (t, 1 H, CH, $J_{H,P}$ = 4.2 Hz), 6.63-6.95 (m, 8 H, m-Ph), 7.13-7.18 (m, 8 H, o-Ph), 7.79-7.84 (m, 4 H, p-Ph) ppm. - ^{13}C$\{^1$H$\}$ NMR (THF-d_8, 100.4 MHz, 25 °C): δ = 3.5 ($SiMe_3$), 16.3 (CH), 128.1, 128.6 (2 t, m-Ph, $J_{C,P}$ = 6 Hz), 130.8 (p-Ph), 131.1 (t, o-Ph, $J_{C,P}$ = 5 Hz), 131.7 (p-Ph), 132.1 (t, o-Ph, $J_{C,P}$ = 5 Hz), 135.8 (i-Ph) ppm. - ^{31}P$\{^1$H$\}$ NMR (THF-d_8, 101.3 MHz, 25 °C): δ = 17.8 (d, $^2J_{P,Y}$ = 7.6 Hz) ppm. - ^{11}B NMR (THF-d_8, 128.15 MHz, 25 °C): δ = -24.7 (br qt, $J_{B,H}$ = 108 Hz) ppm. - IR (v/cm^{-1}): 733 (s), 841 (vs), 987 (m), 1094 (vs), 1192 (m), 1256 (w), 1437 (m), 1579 (w), 1971 (w), 2163 (s), 2216 (s), 2486 (m), 2943 (m), 3058 (w). - EI/MS (70 eV, 180 °C) *m/z* (%): 736 (19), 721 (22), 676 (M$^+$, 3), 661 (M$^+$ - BH_4, 100), 647 (39), 586 (100), 569 (99), 558 (CH(PPh$_2$NSiMe$_3$)$_2$$^+$, 14), 543 (CH(PPh$_2$NSiMe$_3$)$_2$$^+$ - CH_3, 97), 493 (63), 358 (12), 323 ([M$^+$ - 2 BH_4] / 2, 28), 272 (Ph$_2$PNSiMe$_3$, 15), 183 (18), 135 (27), 73 ($SiMe_3$, 33), 43 (16). - (**6**) $C_{31}H_{47}B_2N_2Si_2P_2Y$ (676.37): calcd. C, 55.05, H, 7.00, N, 4.14; found C, 55.54, H, 7.28, N, 3.42 %.

$[\{(Me_3SiNPPh_2)_2CH\}Lu(BH_4)_2]$ (**7**)

THF (25 ml) was condensed at -78 °C onto a mixture of $[Lu(BH_4)_3(THF)_3]$ (653 mg, 1.50 mmol) and $[K\{CH(PPh_2NSiMe_3)_2\}]$ (**1**) (895 mg, 1.50 mmol) and the resulting reaction mixture was stirred for 16 h at 60 °C. The colorless solution was filtered and the solvent evaporated *in vacuo*. Then, toluene (20 ml) was condensed onto the residue. The mixture was heated carefully until the solution became clear. The solution was layered with *n*-pentane and allowed to stand at room temperature to obtain the product as colorless crystals after 72 h. - Yield: 822 mg, 0.94 mmol, 63 %. - ^{1}H NMR (THF-d_8, 400 MHz, 25 °C): δ = 0.35 (s, 18 H, $SiMe_3$), 1.45-2.08 (br, 8 H, BH_4), 1.83 (t, 1 H, CH, $J_{H,P}$ = 3.8 Hz), 7.07-7.08 (m, 4 H, Ph), 7.27-7.34 (m, 6 H, Ph), 7.46-7.52 (m, 6 H, Ph), 7.79-7.81 (m, 4 H, Ph) ppm. - ^{13}C$\{^1$H$\}$ NMR (THF-d_8, 100.4 MHz, 25 °C): δ = 3.3 ($SiMe_3$), 17.6 (t, CH, $J_{C,P}$ = 100 Hz), 128.6-128.9 (m, m-Ph), 130.5 (p-Ph), 131.5-131.7 (m, o-Ph), 132.2-132.4 (m, i-Ph) ppm. - ^{31}P$\{^1$H$\}$ NMR (THF-d_8, 101.3 MHz, 25 °C): δ = 20.9 ppm. -

^{11}B NMR (THF-d$_8$, 128.15 MHz, 25 °C): δ = - 25.8 ppm. - IR (ν/cm^{-1}): 740 (s), 776 (w), 827 (s), 996 (w), 1112 (s), 1164 (w), 1241 (m), 1436 (m), 1587 (w), 1983 (w), 2079 (m), 2225 (w), 2309 (m), 2420 (w), 2654 (w), 2862 (w), 2945 (w). - EI/MS (70 eV, 180 °C) *m/z* (%): 762 (M$^+$, 9), 747 (M$^+$ - BH$_4$, 90), 733 ({(Me$_3$SiNPPh$_2$)$_2$CH}Lu$^+$, 36), 672 (100), 657 (32), 569 (42), 558 (CH(PPh$_2$NSiMe$_3$)$_2^+$, 14), 543 (CH(PPh$_2$NSiMe$_3$)$_2^+$ - CH$_3$, 96), 493 (23), 366 ([M$^+$ - 2 BH$_4$] / 2, 23), 359 ([{(Me$_3$SiNPPh$_2$)$_2$CH}Lu$^+$ - CH$_3$] / 2,17), 272 (Ph$_2$PNSiMe$_3$, 13), 183 (10), 135 (18), 121 (19), 73 (SiMe$_3$, 21), 43 (10). - (**7** + THF) C$_{35}$H$_{55}$B$_2$N$_2$OSi$_2$P$_2$Lu (834.53): calcd. C, 50.37, H, 6.64, N, 3.36; found C, 50.04, H, 6.45, N, 2.89 %.

4.2.2 Synthesis of 2,5-bis{*N*-(2,6-diisopropylphenyl)iminomethyl}pyrrolyl alkali metal complexes

[(DIP$_2$pyr)Li]$_2$ (**9**)

(DIP$_2$pyr)H (**8**) (883 mg, 2.00 mmol) was dissolved in toluene (20 ml) and cooled to -78 °C. *n*BuLi (2.5 M, 0.8 ml, 2.04 mmol) was added slowly and the solution was stirred for 3 h at room temperature. The solvent was evaporated *in vacuo* and the residue was washed with *n*-pentane. The product was dried *in vacuo* giving a pale yellow powder. - Yield: 655 mg, 0.73 mmol, 73 %. Single crystals were obtained by crystallization from hot toluene. - ^{1}H NMR (THF-d$_8$, 400 MHz, 25 °C): δ = 1.18 (d, 24 H, CH(C*H*$_3$), $J_{H,H}$ = 6.9 Hz), 3.22 (sept, 4 H, C*H*(CH$_3$), $J_{H,H}$ = 6.9 Hz), 6.71 (m, 2 H, 3,4-pyr), 6.89-7.03 (m, 2 H, p-Ph), 7.10-7.13 (m, 4 H, Ph), 7.95 (m, 2 H, N=CH) ppm. - ^{1}H NMR (C$_6$D$_6$, 400 MHz, 25 °C):[163, 164] δ = 1.05 (d, 24 H, CH(C*H*$_3$)$_2$, $J_{H,H}$ = 6.4 Hz), 3.16 (sept, 4 H, C*H*(CH$_3$)$_2$, $J_{H,H}$ = 6.4 Hz), 6.90 (s, 2 H, 3,4-pyr), 7.31 (m, 6 H, Ph), 8.18 (s, 2 H, N=CH). - ^{13}C{^{1}H} NMR (THF-d$_8$, 100.4 MHz, 25 °C): δ = 24.1 (CH(CH$_3$)$_2$), 28.1 (CH(CH$_3$)$_2$), 117.2 (3,4-pyr), 123.2 (Ph), 123.7 (Ph), 139.5 (2,5-pyr), 144.3 (Ph), 151.3 (Ph), 160.5 (N=CH) ppm. - ^{7}Li{^{1}H} NMR (C$_6$D$_6$, 400 MHz, 25 °C): 3.6 ppm. - (**9**) C$_{60}$H$_{76}$N$_6$Li$_2$ (895.17): calcd. C, 80.50, H, 8.56, N, 9.39; found C, 80.09, H, 8.54, N, 9.09 %.

[(DIP$_2$pyr)Na]$_2$ (**11**)

THF (40 ml) was added to a mixture of (DIP$_2$pyr)H (**8**) (883 mg, 2.00 mmol) and NaH (53 mg, 2.20 mmol) and the reaction mixture was refluxed for 4 h. After cooling to room temperature the excess of NaH was filtered off and the solvent was removed *in vacuo*. The residue was washed with *n*-pentane and dried *in vacuo*, to obtain the product as pale yellow powder. - Yield: 735 mg, 0.79 mmol, 79 %. Single crystals were obtained by crystallization from hot toluene. - ^{1}H NMR (THF-d$_8$, 400 MHz, 25 °C): δ = 1.22 (d, 24 H, CH(CH_3), $J_{H,H}$ = 6.9 Hz), 3.26 (sept, 4 H, CH(CH$_3$), $J_{H,H}$ = 6.9 Hz), 6.76 (s, 2 H, 3,4-pyr), 6.99-7.04 (m, 2 H, p-Ph), 7.12-7.15 (m, 4 H, Ph), 7.98 (s, 2 H, N=CH) ppm. - ^{13}C{^{1}H} NMR (THF-d$_8$, 100.4 MHz, 25 °C): δ = 23.8 (CH(CH$_3$)$_2$), 28.1 (CH(CH$_3$)$_2$), 117.0 (3,4-pyr), 122.9 (Ph), 123.2 (Ph), 139.2 (2,5-pyr), 144.5 (Ph), 152.4 (Ph), 160.5 (N=CH) ppm. - (**11**) C$_{60}$H$_{76}$N$_6$Na$_2$ (927.27): calcd. C, 77.72, H, 8.26, N, 9.06; found C, 77.71, H, 8.43, N, 8.52 %.

4.2.3 Synthesis of 2,5-bis{*N*-(2,6-diisopropylphenyl)iminomethyl}pyrrolyl rare earth metal chlorides and borohydrides

[(DIP$_2$pyr)NdCl$_2$(THF)]$_2$ (**12**)

THF (10 ml) was condensed onto a mixture of NdCl$_3$ (426 mg, 1.70 mmol) and [(DIP$_2$pyr)K] (**10**) (816 mg, 1.70 mmol) and the mixture was stirred at 60 °C overnight. The resulting yellow solution was filtered off, concentrated and layered with *n*-pentane. The product was obtained as a yellow microcrystalline solid. - Yield 750 mg, 0.26 mmol, 61 %. - ^{1}H NMR (THF-d$_8$, 300 MHz, 25 °C): δ = -3.02 (br, 12 H, CH(CH_3)), -2.59 (br, 12 H, CH(CH_3)), 0.14 (d, 4 H, Ph, $J_{H,H}$ = 7.5 Hz), 0.95 (t, 2 H, p-Ph, $J_{H,H}$ = 7.8 Hz), 16.58 (br, 2 H, 3,4-pyr), 17.83 (br, 2 H, N=CH) ppm. - IR (KBr, ν/cm^{-1}): 858(m), 1055(s), 1099(m), 1162(vs), 1344(s), 1411(m), 1441(s), 1460(s), 1565(vs), 1623(vs), 2866(m), 2926(m), 2960(s). - (**12**) C$_{68}$H$_{92}$N$_6$O$_2$Cl$_4$Nd$_2$ (1455.79): calcd. C, 56.10, H, 6.37, N, 5.77; found C, 55.64, H, 5.92, N, 5.82.

[{DIP$_2$pyr*-BH$_3$}Sc(BH$_4$)(THF)$_2$] (**15**)

THF (20 ml) was condensed at -78 °C onto a mixture of [Sc(BH$_4$)$_3$(THF)$_2$] (306 mg, 1.30 mmol) and [(DIP$_2$pyr)K] (**10**) (480 mg, 1.00 mmol) and the resulting yellow reaction mixture was stirred for 16 h at 60 °C. The yellow solution was filtered off and concentrated until a yellow precipitate appeared. The mixture was heated carefully until the solution became clear. The solution was allowed to stand at room temperature to obtain the product as yellow crystals after several hours. - Yield 527 mg, 0.66 mmol, 66 %. - ^{1}H NMR (THF-d$_8$, 400 MHz, 25 °C): δ = -0.08-0.77 (br, 7 H, BH$_4$, BH$_3$), 1.11-1.31 (m, 24 H, CH(CH_3)), 3.17 (sept, 2 H, CH(CH$_3$), $J_{H,H}$ = 6.8 Hz), 3.72 (sept, 2 H, CH(CH$_3$), $J_{H,H}$ = 6.8 Hz), 4.46 (s, 2 H, N-CH$_2$), 6.24 (d, 1 H, pyr, $J_{H,H}$ = 3.6 Hz), 6.97-6.99 (m, 1 H, Ph) 7.03-7.05 (m, 2 H, Ph), 7.08 (d, 1 H, pyr, $J_{H,H}$ = 3.6 Hz), 7.20-7.27 (m, 3 H, Ph), 7.61 (s, 1 H, N=CH) ppm. - ^{13}C{^{1}H} NMR (THF-d$_8$, 100.4 MHz, 25 °C): δ = 23.8, 24.3, 26.0, 26.2 (CH(CH$_3$)$_2$), 27.1, 28.8 (CH(CH$_3$)$_2$), 59.4 (N-CH$_2$), 109.3 (3,4-pyr), 123.6 (Ph), 124.2 (Ph), 127.6, 129.4, 133.4, 142.7 (2,5-pyr), 143.3, 145.1, 149.3, 153.7, 154.5, 163.6 ppm. -^{11}B-NMR (THF-d$_8$, 128.15 MHz, 25 °C): δ = - 16.5 (br, BH$_3$), -24.6 (br, BH$_4$) ppm.- IR (KBr, v/cm^{-1}): 754 (s), 804 (s), 858 (s), 1022 (m), 1046 (s), 1102 (w), 1132 (w), 1196 (w), 1246 (m), 1283 (m), 1317 (s), 1468 (m), 1583 (s), 1605 (vs), 1716 (w), 1786 (w), 1886 (w), 1994 (w), 2094 (m), 2322 (w), 2467 (w), 2866 (w), 2924 (w), 2962 (m). - (**15** + 2 THF) C$_{46}$H$_{78}$B$_2$N$_3$O$_4$Sc (803.71): calcd. C, 68.74, H, 9.78, N, 5.23; found C, 68.36, H, 9.45, N, 5.54.

[(DIP$_2$pyr)LaClBH$_4$(THF)]$_2$ (**17**)

THF (20 ml) was condensed at -78 °C onto a mixture of [La(BH$_4$)$_3$(THF)$_3$] (200 mg, 0.50 mmol), LaCl$_3$ (123 mg, 0.50 mmol) and [(DIP$_2$pyr)K] (**10**) (480 mg, 1.00 mmol) and the resulting yellow reaction mixture was stirred for 16 h at 60 °C. The yellow solution was concentrated to 10 ml and filtered. The solution was allowed to stand at room temperature to obtain the product as pale yellow rhombic crystals after several hours. - Yield 343 mg, 0.27 mmol, 54 % (single crystals). - IR (v/cm^{-1}): 735 (s), 758 (m), 793 (m), 860 (s), 1012 (m), 1051 (s), 1093 (m), 1159 (s), 1246 (w), 1315 (s), 1335 (s), 1437 (m), 1558 (s), 1651 (w), 1737 (w), 1975 (m), 2031 (m), 2046 (m), 2158 (m), 2210 (m), 2447 (w), 2607 (w), 2860 (w), 2951 (m). - (**17** + 2 THF) C$_{76}$H$_{116}$B$_2$N$_6$O$_4$Cl$_2$La$_2$ (1414.57): calcd. C, 58.96, H, 7.55, N, 5.43; found C, 58.77, H, 7.45, N, 5.35.

[(DIP$_2$pyr)NdClBH$_4$(THF)]$_2$ (**18**)

THF (20 ml) was condensed at -78 °C onto a mixture of [Nd(BH$_4$)$_3$(THF)$_3$] (171 mg, 0.42 mmol), NdCl$_3$ (106 mg, 0.42 mmol) and [(DIP$_2$pyr)K] (**10**) (405 mg, 0.84 mmol) and the resulting yellow reaction mixture was stirred for 16 h at 60 °C. The yellow solution was filtered off and concentrated until a yellow precipitate appears. The mixture was heated carefully until the solution became clear. The solution was allowed to stand at room temperature to obtain the product as yellow crystals after several hours. - Yield 250 mg, 0.18 mmol, 42 % (single crystals). - IR (KBr, v/cm^{-1}): 734 (s), 802 (m), 860 (m), 1012 (m), 1052 (s), 1098 (m), 1163 (s), 1254 (w), 1344 (s), 1440 (m), 1563 (vs), 1731 (w), 1801 (w), 1862 (w), 2048 (m), 2079 (m), 2105 (m), 2229 (w), 2318 (w), 2451 (w), 2654 (w), 2863 (w), 2953 (m). - (**18**) C$_{68}$H$_{100}$B$_2$N$_6$O$_2$Cl$_2$Nd$_2$ (1414.57): calcd. C, 57.74, H, 7.13, N, 5.94; found C, 58.32, H, 7.24, N, 5.56.

4.2.4 Synthesis of 2,5-bis{*N*-(2,6-diisopropylphenyl)iminomethyl}pyrrolyl complexes of divalent lanthanides and alkaline earth metals

[(DIP$_2$pyr)LnI(THF)$_3$] (**19-21**)

THF (20 ml) was condensed at -78 °C onto a mixture of LnI$_2$(THF)$_n$ (Ln = Sm, Eu, Yb) and [(DIP$_2$pyr)K] (**10**) and the resulting reaction mixture was stirred for 16 h at 60 °C. The deeply colored solution was filtered off and concentrated until a precipitate appears. The mixture was heated carefully until the solution became clear. For Ln = Yb the solution was layered with *n*-pentane. The products were obtained at room temperature overnight as deeply colored crystals.

[(DIP$_2$pyr)SmI(THF)$_3$] (**19**):

SmI$_2$(THF)$_{2.5}$ (643 mg, 1.10 mmol), [(DIP$_2$pyr)K] (**10**) (480 mg, 1.00 mmol). - Black crystals from hot THF, yield 706 mg, 0.76 mmol, 76 %.- IR (KBr, v/cm^{-1}): 731 (m), 766 (m), 800 (m), 860 (m), 1039 (m), 1093 (m), 1155 (s), 1254 (w), 1315 (m), 1360 (w), 1437 (m), 1574 (m), 1616 (s), 1714 (w), 2046 (w), 2181 (w), 2864 (w), 2927 (w), 2963 (m), 3450 (m). - (**19** - 0.5 THF) C$_{40}$H$_{58}$N$_3$O$_{2.5}$ISm (935.83): calcd. C, 53.49, H, 6.51, N, 4.68; found C, 53.33, H, 6.63, N, 4.26 %.

[(DIP$_2$pyr)EuI(THF)$_3$] (**20**):

EuI$_2$(THF)$_2$ (412 mg, 0.75 mmol), [(DIP$_2$pyr)K] (**10**) (360 mg, 0.75 mmol). - Red crystals from hot THF, yield 360 mg, 0.39 mmol, 51 % (single crystals). - IR (KBr, v/cm^{-1}): 735 (m), 769 (m), 796 (m), 860 (m), 1039 (m), 1097 (m), 1155 (s), 1230 (w), 1319 (m), 1360 (w), 1437 (m), 1574 (m), 1616 (s), 1710 (w), 2046 (w), 2181 (w), 2868 (w), 2924 (w), 2962 (m), 3450 (m). - (**20**) C$_{42}$H$_{62}$N$_3$O$_3$IEu (935.83): calcd. C, 53.90, H, 6.68, N, 4.49; found C, 53.58, H, 6.70, N, 4.21 %.

[(DIP$_2$pyr)YbI(THF)$_3$] (**21**):

YbI$_2$(THF)$_2$ (286 mg, 0.50 mmol), [(DIP$_2$pyr)K] (**10**) (240 mg, 0.50 mmol). - Green crystals from THF/n-pentane, yield 200 mg, 0.21 mmol, 42 % (single crystals). - ^{1}H NMR (THF-d$_8$, 400 MHz, 25 °C): δ = 1.25 (d, 24 H, CH(CH_3), $J_{H,H}$ = 6.9 Hz), 3.25 (sept, 4 H, CH(CH$_3$), $J_{H,H}$ = 6.9 Hz), 6.75 (s, 2 H, 3,4-pyr), 7.07-7.12 (m, 2 H, p-Ph), 7.17-7.19 (m, 4 H, Ph), 8.21 (s, 2 H, N=CH) ppm. - ^{13}C{^{1}H} NMR (THF-d$_8$, 100.4 MHz, 25 °C): δ = 28.2 (CH(CH_3)$_2$), 34.8 (CH(CH$_3$)$_2$), 118.0 (3,4-pyr), 123.1 (Ph), 124.4 (Ph), 139.7 (Ph), 143.8 (2,5-pyr), 151.4 (Ph), 161.5 (N=CH) ppm. - (**21**) C$_{42}$H$_{62}$N$_3$O$_3$IYb (956.90): calcd. C, 52.72, H, 6.53, N, 4.39; found C, 52.86, H, 6.63, N, 4.23 %.

[(DIP$_2$pyr)MI(THF)$_n$] (**22-24**)

THF (20 ml) was condensed at -78 °C onto a mixture of MI$_2$ (M = Ca, Sr, Ba) (1.00 mmol) and [(DIP$_2$pyr)K] (**10**) (480 mg, 1.00 mmol) and the resulting yellow reaction mixture was stirred at 60 °C for 16 h (M = Ca, 60 h). The yellow solution was filtered off and concentrated until a yellow precipitate appears. The mixture was heated carefully until the solution became clear. The solution was allowed to stand at room temperature to obtain the product as yellow crystals after several hours.

[(DIP$_2$pyr)SrI(THF)$_3$] (**22**):

SrI$_2$ (341 mg, 1.00 mmol). - Yield 634 mg, 0.73 mmol, 73 %. - ^{1}H NMR (THF-d$_8$, 300 MHz, 25 °C): δ = 1.17-1.21 (m, 24 H, CH(C*H$_3$*)), 3.22 (sept, 4 H, C*H*(CH$_3$), $J_{H,H}$ = 6.9 Hz), 6.59 (s, 2 H, 3,4-pyr), 7.05 (dd, 2 H, p-Ph, $J_{H,H,1}$ = 8.8 Hz, $J_{H,H,2}$ = 6.3 Hz), 7.13-7.15 (m, 4 H, Ph), 8.01 (s, 2 H, N=CH) ppm. - ^{13}C{^{1}H} NMR (THF-d$_8$, 100.4 MHz, 25 °C): δ = 25.4 (CH(*CH$_3$*)$_2$), 27.6 (*C*H(CH$_3$)$_2$), 117.3 (3,4-pyr), 122.5 (Ph), 123.8 (Ph), 139.4 (Ph), 142.5 (2,5-pyr), 150.9 (Ph), 161.5 (N=CH) ppm. - (**22** - 0.5 THF) C$_{40}$H$_{58}$N$_3$O$_{2.5}$SrI (835.43): calcd. C, 57.51, H, 7.00, N, 5.03; found C, 56.97, H, 6.98, N, 4.41 %.

[(DIP$_2$pyr)BaI(THF)$_4$] (**23**):

BaI$_2$ (391 mg, 1.00 mmol). - Yield 455 mg, 0.46 mmol, 46 % (single crystals). - ^{1}H NMR (THF-d$_8$, 400 MHz, 25 °C): δ = 1.16-1.32 (m, 24 H, CH(C*H$_3$*)), 3.22 (sept, 4 H, C*H*(CH$_3$), $J_{H,H}$ = 6.9 Hz), 6.59 (s, 2 H, 3,4-pyr), 7.07 (dd, 2 H, p-Ph, $J_{H,H,1}$ = 8.4 Hz, $J_{H,H,2}$ = 6.8 Hz), 7.13-7.17 (m, 4 H, Ph), 8.03 (s, 2 H, N=CH) ppm. - ^{13}C{^{1}H} NMR (THF-d$_8$, 100.4 MHz, 25 °C): δ = 26.1 (CH(*CH$_3$*)$_2$), 28.4 (CH(*CH$_3$*)$_2$), 118.1 (3,4-pyr), 123.3 (Ph), 124.5 (Ph), 140.1 (2,5-pyr), 143.7 (Ph), 151.3 (Ph), 162.4 (N=CH) ppm. - (**23**) C$_{46}$H$_{70}$N$_3$O$_4$IBa (993.30): calcd. C, 55.62, H, 7.10, N, 4.23; found C, 55.21, H, 6.71, N, 4.53 %.

[(DIP$_2$pyr)CaI(THF)$_3$] (**24**):

CaI$_2$ (294 mg, 1.00 mmol). - Yield 505 mg, 0.61 mmol, 61 % (single crystals). - ^{1}H NMR (THF-d$_8$, 300 MHz, 25 °C): δ = 1.20 (d, 24 H, CH(C*H$_3$*), $J_{H,H}$ = 6.9 Hz), 3.22 (sept, 4 H, C*H*(CH$_3$), $J_{H,H}$ = 6.9 Hz), 6.79 (s, 2 H, 3,4-pyr), 7.03 (dd, 2 H, p-Ph, $J_{H,H,1}$ = 8.7 Hz, $J_{H,H,2}$ = 6.6 Hz), 7.11-7.14 (m, 4 H, Ph), 8.24 (br, 2 H, N=CH) ppm. - ^{13}C{^{1}H} NMR (THF-d$_8$, 100.4 MHz, 25 °C): δ = 23.7 (CH(*CH$_3$*)$_2$), 24.1 (br, CH(*CH$_3$*)$_2$), 28.2 (CH(*CH$_3$*)$_2$), 28.3 (*C*H(CH$_3$)$_2$), 116.0 (3,4-pyr), 123.2 (Ph), 123.3 (Ph), 124.3 (br, Ph), 135.0 (Ph), 138.2 (Ph), 139.6 (br, 2,5-pyr), 150.9 (Ph), 153.1 (Ph), 161.4 (br, N=CH) ppm. - (**24**) C$_{42}$H$_{62}$N$_3$O$_3$ICa (823.94): calcd. C, 61.22, H, 7.58, N, 5.10; found C, 60.83, H, 7.69, N, 4.78 %.

Further ^{1}H NMR studies of **24** at high and low temperatures (65 °C and -70 °C):

^{1}H NMR (THF-d$_8$, 300 MHz, 65 °C): δ = 1.19 (d, 24 H, CH(C*H$_3$*), $J_{H,H}$ = 6.9 Hz), 3.17 (sept, 4 H, C*H*(CH$_3$), $J_{H,H}$ = 6.9 Hz), 6.80 (s, 2 H, 3,4-pyr), 7.01 (dd, 2 H, p-Ph, $J_{H,H,1}$ = 8.4 Hz, $J_{H,H,2}$ = 6.6 Hz), 7.09-7.12 (m, 4 H, Ph), 8.23 (s, 2 H, N=CH) ppm. - ^{1}H NMR (THF-d$_8$, 300 MHz, -70 °C): δ = 1.09-1.28 (m, 24 H, CH(C*H$_3$*), 3.03-3.18 (m, 4 H, C*H*(CH$_3$)), 6.59 (s, 2 H, 3,4-pyr), 7.00-7.16 (m, 6 H, Ph), 8.07 (s, 2 H, N=CH) ppm.

[(DIP$_2$pyr)M{N(SiMe$_3$)$_2$}(THF)$_2$] (**25, 26**)

THF (15 ml) was condensed at -78 °C onto a mixture of [(DIP$_2$pyr)MI(THF)$_3$] (M = Ca (**24**), Sr (**22**)) (0.5 mmol) and [K{N(SiMe$_3$)$_2$}] (100 mg, 0.5 mmol) and the resulting yellow reaction mixture was stirred at 60 °C for 16 h. The yellow solution was filtered off and concentrated until a yellow precipitate appears. The mixture was heated carefully until the solution became clear. The solution was allowed to stand at room temperature to obtain the product as yellow crystals after several hours.

[(DIP$_2$pyr)Ca{N(SiMe$_3$)$_2$}(THF)$_2$] (**25**):

[(DIP$_2$pyr)CaI(THF)$_3$] (**24**) (411 mg, 0.5 mmol). - Yield 270 mg, 0.34 mmol, 69 % (single crystals). - ^{1}H NMR (THF-d$_8$, 300 MHz, 25 °C): δ = -0.21 (s, 18 H, SiMe$_3$), 1.21 (d, 24 H, CH(C*H*$_3$), $J_{H,H}$ = 6.9 Hz), 3.07 (sept, 4 H, C*H*(CH$_3$), $J_{H,H}$ = 6.9 Hz), 6.71 (s, 2 H, 3,4-pyr), 7.07 (dd, 2 H, p-Ph, $J_{H,H,1}$ = 8.7 Hz, $J_{H,H,2}$ = 6.0 Hz), 7.13-7.16 (m, 4 H, Ph), 8.07 (s, 2 H, N=CH) ppm. - ^{13}C{^{1}H} NMR (THF-d$_8$, 100.4 MHz, 25 °C): δ = 5.0 (SiMe$_3$), 25.4 (CH(*C*H$_3$)$_2$), 27.7 (*C*H(CH$_3$)$_2$), 118.8 (3,4-pyr), 122.6 (Ph), 124.0 (Ph), 140.0 (Ph), 142.5 (2,5-pyr), 150.5 (Ph), 161.3 (N=CH) ppm. - (**25** + THF) C$_{48}$H$_{80}$N$_4$O$_3$Si$_2$Ca (857.42): calcd. C, 67.24, H, 9.40, N, 6.53; found C, 67.36, H, 9.11, N, 6.32 %.

[(DIP$_2$pyr)Sr{N(SiMe$_3$)$_2$}(THF)$_2$] (**26**):

[(DIP$_2$pyr)SrI(THF)$_3$] (**22**) (436 mg, 0.5 mmol). - Yield 300 mg, 0.36 mmol, 72 % (single crystals). - ^{1}H NMR (THF-d$_8$, 300 MHz, 25 °C): δ = -0.29 (s, 18 H, SiMe$_3$), 1.15-1.32 (m, 24 H, CH(C*H*$_3$)), 3.06 (sept, 4 H, C*H*(CH$_3$), $J_{H,H}$ = 6.9 Hz), 6.62 (s, 2 H, 3,4-pyr), 7.07 (dd, 2 H, p-Ph, $J_{H,H,1}$ = 9.0 Hz, $J_{H,H,2}$ = 6.0 Hz), 7.13-7.16 (m, 4 H, Ph), 8.03 (s, 2 H, N=CH) ppm. - ^{13}C{^{1}H} NMR (THF-d$_8$, 100.4 MHz, 25 °C): δ = 5.0 (SiMe$_3$), 25.4 (CH(*C*H$_3$)$_2$), 27.7 (*C*H(CH$_3$)$_2$), 118.1 (3,4-pyr), 122.6 (Ph), 124.0 (Ph), 139.0 (Ph), 142.5 (2,5-pyr), 150.8 (Ph), 162.0 (N=CH) ppm. - (**26** + THF) C$_{48}$H$_{80}$N$_4$O$_3$Si$_2$Sr (904.96): calcd. C, 63.71, H, 8.91, N, 6.19; found C, 64.30, H, 8.58, N, 6.23 %.

4.3   Hydroamination studies

The catalyst was weighed under argon gas into an NMR tube. C_6D_6 ($\approx$ 0.5 ml) was condensed into the NMR tube, and the mixture was frozen to -196 °C. The reactant was injected onto the solid mixture, and the whole sample was melted and mixed just before insertion into the core of the NMR machine (t_0). The ratio between the reactant and the product was calculated by comparison of the integrations of the corresponding signals. $SiMe_4$ or ferrocene was used as an internal standard for the kinetic measurements. The substrates 2,2-diphenyl-pent-4-enylamine (**27a**),[270] C-(1-allyl-cyclohexyl)-methylamine (**28a**),[270] 2,2-dimethyl-pent-4-enylamine (**29a**),[270] 2-amino-5-hexene (*via* hex-5-en-2-one oxime[271]) (**30a**)[82] and [1-(pent-2-ynyl)cyclohexyl]methanamine (**31a**)[156] were synthesized according to the literature procedures. ^{1}H NMR spectra of 2-methyl-4,4-diphenylpyrrolidine (**27b**),[270] 3-methyl-2-aza-spiro[4.5]decane (**28b**),[270] 2,4,4-trimethylpyrrolidine (**29b**),[270] 2,5-dimethylpyrrolidine (**30b**)[82] and 3-propyl-2-azaspiro[4.5]dec-2-ene (**31b**)[156] conform with the literature.

<u>Isomerization and tautomerism products, *in situ* ^{1}H NMR data:</u>

2-amino-4-hexene (**30c**)
^{1}H NMR (C_6D_6, 300 MHz, 25 °C): δ = 0.73 (br, 2 H, NH_2), 0.92 (d, 3 H, Me, $J_{H,H}$ = 6.3 Hz), 1.47-1.50 (m, 3 H, Me-C=C), 1.92 (t, 2 H, C=C-CH_2, $J_{H,H}$ = 6.9 Hz), 2.71 (m$_s$, 1 H, CH), 5.28-5.38, 5.44-5.54 (2 m, 2 H, HC=CH) ppm.

3-propylidene-2-azaspiro[4.5]decane (**31c**)
^{1}H NMR (C_6D_6, 400 MHz, 25 °C): δ = 0.90 (t, 3 H, Me, $J_{H,H}$ = 7.2 Hz)[a], 1.23-1.36 (m, 10 H, Cy)[a], 1.68-1.72 (m, 2 H, C-CH_2-C=), 2.02 (m, 2 H, C=CH-CH_2-Me)[a], 2.52 (m, 2 H, C-CH_2-NH), 5.43 (m$_s$, 1 H, C=CH) ppm.
[a]signals obscured by other signals

4.4   Crystal structures

4.4.1 General considerations

A suitable crystal of each compound was covered in mineral oil (Aldrich) and mounted onto a glass fiber. The crystal was transferred directly to the -73 °C or -123 N_2 cold stream of a Stoe IPDS 2 or Stoe IPDS 2T diffractometer (MoK$_\alpha$ radiation; λ = 0.71073 Å). Subsequent computations were carried out on an Intel Pentium IV PC or on a Core2Duo.

The low temperature measurement was performed by G. Eikerling, using a MAR345 imaging plate detector system (MARRESEARCH) mounted on a four-circle goniometer with Eulerian geometry (HUBER) with a rotating anode generator (Bruker FR591, MoK$_\alpha$ radiation, λ = 0.71073 Å). The crystal was cooled to 9 K employing a closed-cycle helium cryostat (ARS-4K).

All structures were solved by the direct or the Patterson method (SHELXS-97[272]). The remaining non-hydrogen atoms were located from successive difference Fourier map calculations. The refinements were carried out by using full-matrix least-squares techniques on F, minimizing the function $(F_o\text{-}F_c)^2$, where the weight is defined as $4F_0{}^2/2(F_o{}^2)$ and F_o and F_c are the observed and calculated structure factor amplitudes using the program SHELXL-97.[272] Carbon-bound hydrogen atom positions were calculated and allowed to ride on the carbon to which they are bonded. The hydrogen atom contributions were calculated, but not refined. The locations of the largest peaks in the final difference Fourier map calculation as well as the magnitude of the residual electron densities in each case were of no chemical significance.

4.4.2 Crystallographic data

[{(Me$_3$SiNPPh$_2$)$_2$CH}La(BH$_4$)$_2$(THF)] (**3**)

Chemical formula	C$_{35}$H$_{55}$B$_2$LaN$_2$OP$_2$Si$_2$•2(C$_4$H$_8$O)
Crystal system, Space group	Monoclinic, $P2_1/n$
a/Å, b/Å, c/Å	10.0940(4), 35.3449(16), 14.2018(6)
β/°	103.779(3)
Unit cell volume/Å^3	4921.0(4)
Temperature/K	200(2)
No. of formula units per unit cell, Z	4
Absorption correction	Integration
Absorption coefficient, μ/mm^{-1}	1.018
No. of reflections measured	25833
No. of independent reflections	8647
R_{int}	0.0433
Final R_1 values ($I > 2\sigma(I)$)	0.0408
Final wR_2 values (all data)	0.1135
Goodness of fit	1.038

[{(Me$_3$SiNPPh$_2$)$_2$CH}Nd(BH$_4$)$_2$(THF)] (**4**)

Chemical formula	C$_{35}$H$_{55}$B$_2$N$_2$NdOP$_2$Si$_2$•C$_4$H$_8$O
Crystal system, Space group	Triclinic, P-1
a/Å, b/Å, c/Å	10.0240(11), 14.0591(18), 16.601(3)
α/°, β/°, γ/°	100.341(11), 92.568(10), 104.573(10)
Unit cell volume/Å^3	2217.5(5)
Temperature/K	200(2)
No. of formula units per unit cell, Z	2
Absorption correction	Integration
Absorption coefficient, μ/mm^{-1}	1.330
No. of reflections measured	16352
No. of independent reflections	7780
R_{int}	0.0256
Final R_1 values ($I > 2\sigma(I)$)	0.0211
Final wR_2 values (all data)	0.0529
Goodness of fit	1.035

$[\{(Me_3SiNPPh_2)_2CH\}Sc(BH_4)_2]$ (**5**), T = 200 K

Chemical formula	$C_{31}H_{47}B_2N_2P_2ScSi_2 \cdot C_7H_8$
Crystal system, Space group	Monoclinic, $P2_1$
$a/Å$, $b/Å$, $c/Å$	10.744(2), 11.264(2), 18.065(4)
$\beta/°$	94.20(3)
Unit cell volume/$Å^3$	2180.4(8)
Temperature/K	200(2)
No. of formula units per unit cell, Z	2
Absorption correction	Integration
Absorption coefficient, μ/mm^{-1}	0.323
No. of reflections measured	19026
No. of independent reflections	10295
R_{int}	0.0764
Final R_1 values ($I > 2\sigma(I)$)	0.0589
Final wR_2 values (all data)	0.1365
Goodness of fit	1.039

$[\{(Me_3SiNPPh_2)_2CH\}Sc(BH_4)_2]$ (**5**), T = 9 K

Chemical formula	$C_{31}H_{47}B_2N_2P_2ScSi_2 \cdot C_7H_8$
Crystal system, Space group	Monoclinic, $P2_1/c$
$a/Å$, $b/Å$, $c/Å$	14.6250(5), 12.7346(3), 22.3686(7)
$\beta/°$	102.939(2)
Unit cell volume/$Å^3$	4060.2(2)
Temperature/K	9
No. of formula units per unit cell, Z	4
Absorption coefficient, μ/mm^{-1}	0.347
Absorption correction	Empirical
No. of reflections measured	14577
No. of independent reflections	7043
R_{int}	0.0247
Final R_1 values ($I > 2\sigma(I)$)	0.0438
Final wR_2 values (all data)	0.0838
Goodness of fit	1.202

low data quality

$[\{(Me_3SiNPPh_2)_2CH\}Y(BH_4)_2]$ (**6**)

Chemical formula	$C_{31}H_{47}B_2N_2P_2Si_2Y$
Crystal system, Space group	Monoclinic, $P2_1/n$
$a/Å$, $b/Å$, $c/Å$	9.7964(7), 17.4944(8), 21.4015(15)
$\beta/°$	91.599(6)
Unit cell volume/$Å^3$	3666.4(4)
Temperature/K	200(2)
No. of formula units per unit cell, Z	4
Absorption correction	Integration
Absorption coefficient, μ/mm^{-1}	1.765
No. of reflections measured	17812
No. of independent reflections	6440
R_{int}	0.0614
Final R_1 values ($I > 2\sigma(I)$)	0.0404
Final wR_2 values (all data)	0.0793
Goodness of fit	1.015

$[\{(Me_3SiNPPh_2)_2CH\}Lu(BH_4)_2]$ (**7**)

Chemical formula	$C_{31}H_{47}B_2N_2P_2Si_2Lu$
Crystal system, Space group	Monoclinic, $P2_1/n$
$a/Å$, $b/Å$, $c/Å$	9.7538(6), 17.4791(7), 21.3528(13)
$\beta/°$	91.622(5)
Unit cell volume/$Å^3$	3638.9(3)
Temperature/K	200(2)
No. of formula units per unit cell, Z	4
Absorption correction	Integration
Absorption coefficient, μ/mm^{-1}	2.889
No. of reflections measured	17674
No. of independent reflections	6400
R_{int}	0.0484
Final R_1 values ($I > 2\sigma(I)$)	0.0271
Final wR_2 values (all data)	0.0592
Goodness of fit	1.003

[(DIP$_2$pyr)Li]$_2$ (**9**)

Chemical formula	$C_{60}H_{76}Li_2N_6$
Crystal system, Space group	Monoclinic, $C2/c$
$a/Å$, $b/Å$, $c/Å$	77.222(15), 15.608(3), 18.684(4)
$\beta/°$	96.85(3)
Unit cell volume/$Å^3$	22359(8)
Temperature/K	200(2)
No. of formula units per unit cell, Z	16
Absorption correction	None
Absorption coefficient, μ/mm^{-1}	0.062
No. of reflections measured	45705
No. of independent reflections	18020
R_{int}	0.0924
Final R_1 values ($I > 2\sigma(I)$)	0.0910
Final wR_2 values (all data)	0.2499
Goodness of fit	0.995

[(DIP$_2$pyr)Na]$_2$ (**11**)

Chemical formula	$C_{60}H_{76}N_6Na_2 \cdot 1.5(C_7H_8)$
Crystal system, Space group	Triclinic, P-1
$a/Å$, $b/Å$, $c/Å$	11.205(2), 12.304(3), 24.047(5)
$\alpha/°$, $\beta/°$, $\gamma/°$	83.20(3), 80.33(3), 80.59(3)
Unit cell volume/$Å^3$	3210.2(11)
Temperature/K	200(2)
No. of formula units per unit cell, Z	2
Absorption correction	None
Absorption coefficient, μ/mm^{-1}	0.076
No. of reflections measured	26156
No. of independent reflections	12534
R_{int}	0.0724
Final R_1 values ($I > 2\sigma(I)$)	0.0699
Final wR_2 values (all data)	0.1976
Goodness of fit	1.003

[(DIP$_2$pyr)NdCl$_2$(THF)]$_2$ (**12**)

Chemical formula	C$_{38}$H$_{54}$N$_3$O$_2$Cl$_2$Nd
Crystal system, Space group	Triclinic, *P*-1
a/Å, *b*/Å, *c*/Å	10.9550(5), 14.3112(7), 14.7917(7)
α/°, *β*/°, *γ*/°	65.656(3), 68.791(4), 89.498(4)
Unit cell volume/Å^3	1942.65(16)
Temperature/K	150(2)
No. of formula units per unit cell, *Z*	2
Absorption correction	None
Absorption coefficient, *μ*/mm^{-1}	1.508
No. of reflections measured	13422
No. of independent reflections	7270
R$_{int}$	0.1249
Final *R*$_1$ values (*I* > 2*σ*(*I*))	0.0805
Final *wR*$_2$ values (all data)	0.2131
Goodness of fit	1.094

[{DIP$_2$pyr*-BH$_3$}Sc(BH$_4$)(THF)$_2$] (**15**)

Chemical formula	C$_{38}$H$_{62}$B$_2$N$_3$O$_2$Sc•2(C$_4$H$_8$O)
Crystal system, Space group	Monoclinic, *Pn*
a/Å, *b*/Å, *c*/Å	9.7039(4), 15.4721(7), 16.5184(7)
β/°	100.561(3)
Unit cell volume/Å^3	2438.06(18)
Temperature/K	200(2)
No. of formula units per unit cell, *Z*	2
Absorption coefficient, *μ*/mm^{-1}	0.192
Absorption correction	Integration
No. of reflections measured	17782
No. of independent reflections	9120
R$_{int}$	0.0543
Final *R*$_1$ values (*I* > 2*σ*(*I*))	0.0756
Final *wR*$_2$ values (all data)	0.2169
Goodness of fit	1.040

[(DIP$_2$pyr)LaClBH$_4$(THF)]$_2$ (**17**)

Chemical formula	$C_{68}H_{100}B_2N_6O_2Cl_2La_2 \cdot 5(C_4H_8O)$
Crystal system, Space group	Monoclinic, $C2/c$
a/Å, b/Å, c/Å	29.055(6), 14.167(3), 23.347(5)
β/°	106.16(3)
Unit cell volume/Å^3	9230(3)
Temperature/K	150(2)
No. of formula units per unit cell, Z	4
Absorption correction	Integration
Absorption coefficient, μ/mm^{-1}	1.023
No. of reflections measured	63716
No. of independent reflections	9800
R_{int}	0.0712
Final R_1 values ($I > 2\sigma(I)$)	0.0402
Final wR_2 values (all data)	0.1103
Goodness of fit	1.051

[(DIP$_2$pyr)NdClBH$_4$(THF)]$_2$ (**18**)

Chemical formula	$C_{68}H_{100}B_2N_6O_2Cl_2Nd_2 \cdot 5(C_4H_8O)$
Crystal system, Space group	Monoclinic, $C2/c$
a/Å, b/Å, c/Å	29.057(6), 14.147(3), 23.195(5)
β/°	106.47(3)
Unit cell volume/Å^3	9144(3)
Temperature/K	150(2)
No. of formula units per unit cell, Z	4
Absorption correction	Integration
Absorption coefficient, μ/mm^{-1}	1.234
No. of reflections measured	31749
No. of independent reflections	10928
R_{int}	0.0379
Final R_1 values ($I > 2\sigma(I)$)	0.0417
Final wR_2 values (all data)	0.1106
Goodness of fit	1.034

$[(DIP_2pyr)SmI(THF)_3]$ (**19**)

Chemical formula	$C_{42}H_{62}N_3O_3ISm \cdot 2(C_4H_8O)$
Crystal system, Space group	Monoclinic, $P2_1/c$
a/Å, b/Å, c/Å	15.174(3), 31.689(6), 11.265(2)
β/°	108.06(3)
Unit cell volume/Å^3	5149.9(18)
Temperature/K	150(2)
No. of formula units per unit cell, Z	4
Absorption correction	Integration
Absorption coefficient, μ/mm^{-1}	1.784
No. of reflections measured	10921
No. of independent reflections	10921
R_{int}	0.0000
Final R_1 values ($I > 2\sigma(I)$)	0.0504
Final wR_2 values (all data)	0.1092
Goodness of fit	0.745

$[(DIP_2pyr)EuI(THF)_3]$ (**20**)

Chemical formula	$C_{42}H_{62}EuN_3O_3I \cdot 2(C_4H_8O)$
Crystal system, Space group	Monoclinic, $P2_1/c$
a/Å, b/Å, c/Å	15.174(3), 31.689(6), 11.265(2)
β/°	108.06(3)
Unit cell volume/Å^3	5149.9(18)
Temperature/K	200(2)
No. of formula units per unit cell, Z	4
Absorption correction	Integration
Absorption coefficient, μ/mm^{-1}	1.862
No. of reflections measured	36611
No. of independent reflections	9904
R_{int}	0.0980
Final R_1 values ($I > 2\sigma(I)$)	0.0782
Final wR_2 values (all data)	0.1856
Goodness of fit	1.063

[(DIP$_2$pyr)YbI(THF)$_3$] (**21**)

Chemical formula	C$_{42}$H$_{62}$N$_3$O$_3$IYb•0.5(C$_5$H$_{12}$)
Crystal system, Space group	Monoclinic, $P2_1/c$
a/Å, b/Å, c/Å	17.960(4), 14.767(3), 17.885(4)
β/°	103.47(3)
Unit cell volume/Å^3	4612.9(16)
Temperature/K	150(2)
No. of formula units per unit cell, Z	4
Absorption correction	Integration
Absorption coefficient, μ/mm^{-1}	2.735
No. of reflections measured	12163
No. of independent reflections	6344
R_{int}	0.1316
Final R_1 values ($I > 2\sigma(I)$)	0.0776
Final wR_2 values (all data)	0.1875
Goodness of fit	0.853

low data quality

[(DIP$_2$pyr)SrI(THF)$_3$] (**22**)

Chemical formula	C$_{42}$H$_{62}$N$_3$O$_3$SrI•2(C$_4$H$_8$O)
Crystal system, Space group	Monoclinic, $p2_1/c$
a/Å, b/Å, c/Å	15.269(3), 32.008(6), 11.338(2)
β/°	107.78(3)
Unit cell volume/Å^3	5276.3(18)
Temperature/K	200(2)
No. of formula units per unit cell, Z	4
Absorption correction	None
Absorption coefficient, μ/mm^{-1}	1.651
No. of reflections measured	23303
No. of independent reflections	6036
R_{int}	0.0843
Final R_1 values ($I > 2\sigma(I)$)	0.0563
Final wR_2 values (all data)	0.1523
Goodness of fit	1.020

low data quality

[(DIP$_2$pyr)BaI(THF)$_4$] (**23**)

Chemical formula	2(C$_{46}$H$_{70}$N$_3$O$_4$IBa)•2(C$_4$H$_8$O)
Crystal system, Space group	Monoclinic, $P2_1$
a/Å, b/Å, c/Å	10.979(2), 31.676(6), 15.769(3)
β/°	109.07(3)
Unit cell volume/Å^3	5183.1(18)
Temperature/K	200(2)
No. of formula units per unit cell, Z	2
Absorption correction	Integration
Absorption coefficient, μ/mm^{-1}	1.406
No. of reflections measured	39822
No. of independent reflections	23278
R_{int}	0.0492
Final R_1 values ($I > 2\sigma(I)$)	0.0397
Final wR_2 values (all data)	0.0993
Goodness of fit	1.007

[(DIP$_2$pyr)CaI(THF)$_3$] (**24**)

Chemical formula	C$_{42}$H$_{62}$N$_3$O$_3$CaI•C$_4$H$_8$O
Crystal system, Space group	Monoclinic, $P2_1/c$
a/Å, b/Å, c/Å	17.187(3), 15.177(3), 19.291(4)
β/°	107.76(3)
Unit cell volume/Å^3	4792.4(17)
Temperature/K	150(2)
No. of formula units per unit cell, Z	4
Absorption correction	None
Absorption coefficient, μ/mm^{-1}	0.816
No. of reflections measured	30443
No. of independent reflections	8441
R_{int}	0.1016
Final R_1 values ($I > 2\sigma(I)$)	0.0729
Final wR_2 values (all data)	0.1963
Goodness of fit	0.986

[(DIP$_2$pyr)Ca{N(SiMe$_3$)$_2$}(THF)$_2$] (**25**)

Chemical formula	$C_{44}H_{72}N_4O_2Si_2Ca \cdot C_4H_8O$
Crystal system, Space group	Monoclinic, $P2_1/c$
a/Å, b/Å, c/Å	16.426(3), 10.568(2), 29.727(6)
β/°	97.67(3)
Unit cell volume/Å^3	5114.2(18)
Temperature/K	150(2)
No. of formula units per unit cell, Z	4
Absorption correction	Integration
Absorption coefficient, μ/mm^{-1}	0.210
No. of reflections measured	22279
No. of independent reflections	10661
R_{int}	0.0614
Final R_1 values ($I > 2\sigma(I)$)	0.0689
Final wR_2 values (all data)	0.1790
Goodness of fit	0.970

[(DIP$_2$pyr)Sr{N(SiMe$_3$)$_2$}(THF)$_2$] (**26**)

Chemical formula	$C_{44}H_{72}N_4O_2Si_2Sr \cdot C_4H_8O$
Crystal system, Space group	Monoclinic, $P2_1/c$
a/Å, b/Å, c/Å	16.605(3), 10.727(2), 29.874(6)
β/°	96.87(3)
Unit cell volume/Å^3	5282.8(18)
Temperature/K	200(2)
No. of formula units per unit cell, Z	4
Absorption correction	Integration
Absorption coefficient, μ/mm^{-1}	1.142
No. of reflections measured	39441
No. of independent reflections	10530
R_{int}	0.0968
Final R_1 values ($I > 2\sigma(I)$)	0.0656
Final wR_2 values (all data)	0.1584
Goodness of fit	1.057

5.   Summary / Zusammenfassung

5.1   Summary

Bis(phosphinimino)methanide rare earth metal bisborohydrides, as illustrated in Scheme I, were successfully synthesized by salt metathesis reactions of $[K\{CH(PPh_2NSiMe_3)_2\}]$ with $[Ln(BH_4)_3(THF)_n]$ (Ln = Sc (n = 2); Ln = La, Nd, Lu (n = 3)) or in the case of yttrium by the reaction of $[\{(Me_3SiNPPh_2)_2CH\}YCl_2]_2$ with $NaBH_4$. Interestingly, the BH_4^- anions are η^3-coordinated in the solid state structures of **3**, **4**, **6** and **7**, while for the scandium complex **5** two different conformational polymorphs were identified, in which either both BH_4^- groups are η^3-coordinated or one BH_4^- anion shows an η^2-coordination mode. Furthermore, complexes **3**, **6** and **7** showed high activities in the ring-opening polymerization (ROP) of ε-caprolactone (CL). At 0 °C, the molar mass distribution reached the narrowest values ever obtained for the ROP of CL initiated by a rare earth metal borohydride species.

Ln = La (**3**), Nd (**4**)   Ln = Sc (**5**), Y (**6**), Lu (**7**)

Scheme I

In collaboration with N. Meyer, rare earth metal chlorides and borohydrides of the 2,5-bis$\{N$-(2,6-diisopropylphenyl)iminomethyl$\}$pyrrolyl ligand were synthesized, as shown in Scheme II. The reaction of $[(DIP_2pyr)K]$ (**10**) with anhydrous neodymium trichloride afforded $[(DIP_2pyr)NdCl_2(THF)]_2$ (**12**) which is dimeric in the solid state. Excitingly, the reaction of $[(DIP_2pyr)K]$ (**10**) with $[Ln(BH_4)_3(THF)_n]$ (Ln = Sc (n = 2); Ln = La, Nd, Lu (n = 3)) depends on the ionic radii of the center metals. For the larger rare earth metals lanthanum and neodymium, the expected products $[(DIP_2pyr)Ln(BH_4)_2(THF)_2]$ (Ln = La (**13**), Nd (**14**)) were obtained; while for the smaller rare earth metals scandium and lutetium, an unusual redox reaction of a BH_4^- anion with one of the Schiff-base functions of the ligand was observed and the products $[\{DIP_2pyr^*-BH_3\}Ln(BH_4)(THF)_2]$ (Ln = Sc (**15**), Lu (**16**)) were formed (Scheme II). Moreover, the two neodymium containing complexes **12** and **14** were investigated as Ziegler-Natta catalysts for the polymerization of 1,3-butadiene to form poly-*cis*-1,4-butadiene, by using various cocatalyst mixtures. Very high activities and good selectivities were observed for **12**.

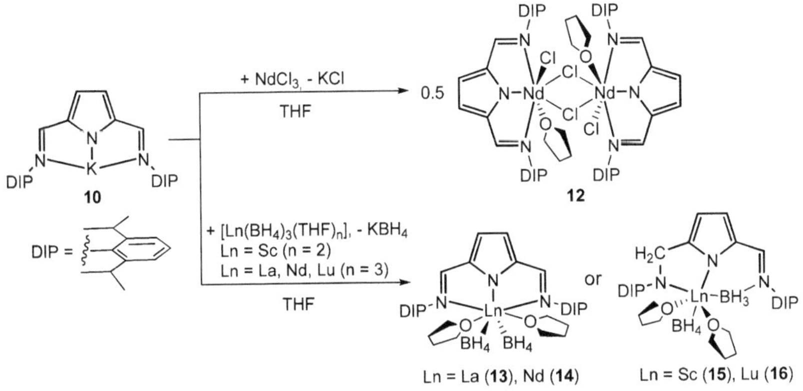

Scheme II

The 2,5-bis{*N*-(2,6-diisopropylphenyl)iminomethyl}pyrrolyl ligand was successfully introduced into the coordination chemistry of the divalent lanthanides and the alkaline earth metals. As shown in Scheme III, salt metathesis reactions of [(DIP$_2$pyr)K] (**10**) with either anhydrous lanthanide diiodides or alkaline earth metal diiodides afforded the corresponding heteroleptic iodo complexes [(DIP$_2$pyr)LnI(THF)$_3$] (Ln = Sm (**19**), Eu (**20**), Yb (**21**)) or [(DIP$_2$pyr)MI(THF)$_n$] (M = Ca (**24**), Sr (**22**) (n = 3); Ba (**23**) (n = 4)). Surprisingly, all complexes **19-24** are monomeric in the solid state, independently from the ionic radii of their center metals. Instead of forming dimers, the coordination sphere of each metal center is satisfied by additionally coordinated THF molecules, which is a very rare structural motif in the chemistry of the larger divalent lanthanides and alkaline earth metals. While the (DIP$_2$pyr)$^-$ ligands in **19-23** are η^3-coordinated in the solid state, for the calcium complex **24** an η^2-coordination mode was observed (Scheme III). Interestingly, the calcium complex **24** and the analogous ytterbium compound **21** show different structures in the solid state.

In order to obtain catalytically active species, [(DIP$_2$pyr)M{N(SiMe$_3$)$_2$}(THF)$_2$] (M = Ca (**25**), Sr (**26**)) were prepared by the reaction of [(DIP$_2$pyr)MI(THF)$_3$] (M = Ca (**24**), Sr (**22**)) with [K{N(SiMe$_3$)$_2$}] (Scheme IV). Compounds **25** and **26** were investigated for the intramolecular hydroamination of aminoalkenes and one aminoalkyne. Unfortunately, both catalysts exhibit a limited reaction scope, caused by the formation of undesired side products by alkene isomerization and imine-enamine tautomerism. However, both compounds are active catalysts and show high yields and short reaction times. The highest activities were observed for the calcium complex **25** and can be compared to the results obtained with the β-diketiminato calcium amide [{(DIPNC(Me))$_2$CH}Ca{N(SiMe$_3$)$_2$}(THF)] as a catalyst.

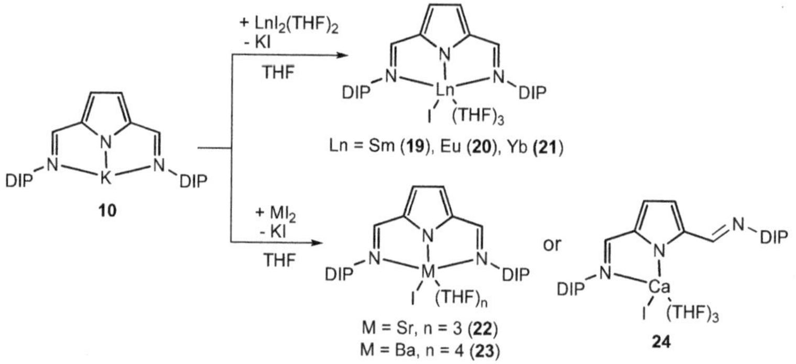
+ LnI2(THF)2
- KI
THF
Ln = Sm (19), Eu (20), Yb (21)
10
+ MI2
- KI
THF
or
M = Sr, n = 3 (22)
M = Ba, n = 4 (23)
24

Scheme III

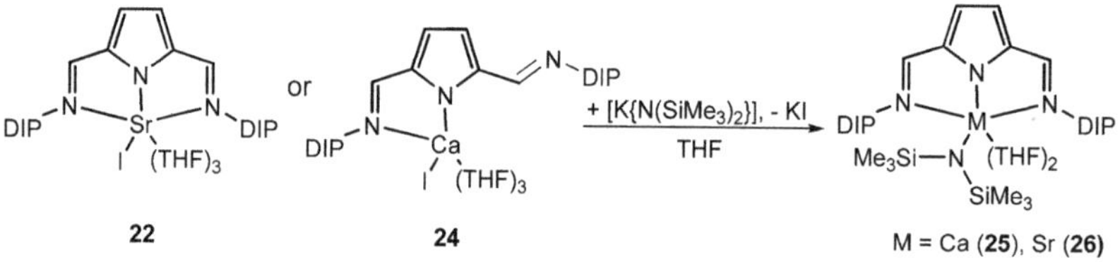
or
+ [K{N(SiMe3)2}], - KI
THF
22
24
M = Ca (25), Sr (26)

Scheme IV

Finally, imidazolin-2-imide and cyclopentadienyl-imidazolin-2-imine rare earth metal alkyl complexes, synthesized by M. Tamm *et al.*, were investigated for the intramolecular hydroamination of non-activated aminoalkenes and one aminoalkyne. Both compounds showed high selectivities and activities, and although they cannot compete with the metallocene analogues, the imidazolin-2-imide complexes are new and interesting examples for catalytically active post-metallocenes.

5.2   Zusammenfassung

Die in Schema I dargestellten Bisborhydridokomplexe der Seltenerdmetalle mit dem Bis(phosphinimino)methanid-Liganden konnten ausgehend von [K{CH(PPh$_2$NSiMe$_3$)$_2$}], in Salzmetathesereaktionen mit [Ln(BH$_4$)$_3$(THF)$_n$] (Ln = Sc (n = 2); Ln = La, Nd, Lu (n = 3)), oder im Falle des Yttriums, durch die Reaktion von [{(Me$_3$SiNPPh$_2$)$_2$CH}YCl$_2$]$_2$ mit NaBH$_4$ synthetisiert werden. Die Borhydrido-Liganden der Komplexe **3**, **4**, **6** und **7** weisen einen η^3-Koordinationsmodus im Festkörper auf, wohingegen interessanterweise für den polymorphen Scandiumkomplex **5** zwei verschiedene Strukturen nachgewiesen werden konnten, in denen entweder beide Borhydridanionen η^3-koordiniert sind, oder ein Borhydrido-Ligand eine η^2-Koordination aufweist. Desweiteren erwiesen sich die Komplexe **3**, **6** und **7** als aktive Katalysatoren für die Ringöffnungspolymerisation von ε-Caprolacton, mit sehr hohen Selektivitäten bei 0 °C. Tatsächlich wurden bei dieser Temperatur die niedrigsten Werte für die Molekularmassenverteilung erreicht, die jemals in der Seltenerdmetallborhydrid-katalysierten Ringöffnungspolymerisation von ε-Caprolacton beobachtet wurden.

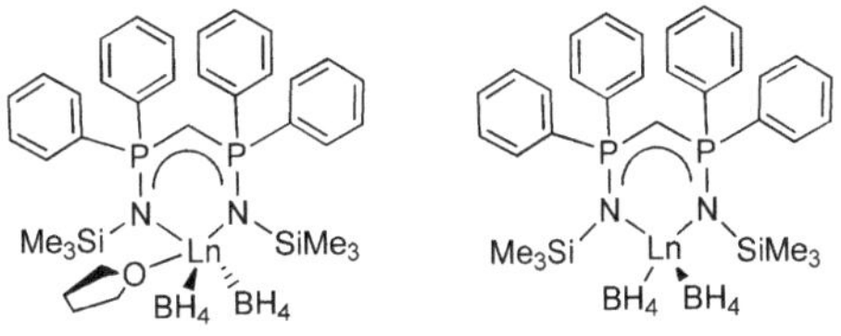

Schema I

In Zusammenarbeit mit N. Meyer wurden Borhydrido- und Chlorokomplexe der Seltenerdmetalle mit dem 2,5-bis{N-(2,6-diisopropylphenyl)iminomethyl}pyrrolyl-Liganden synthetisiert. Die Darstellung des dimeren Neodymkomplexes [(DIP$_2$pyr)NdCl$_2$(THF)]$_2$ (**12**) erfolgte durch den Umsatz von [(DIP$_2$pyr)K] (**10**) mit wasserfreiem Neodymtrichlorid (Schema II). Die Reaktion des Kaliumsalzes **10** mit [Ln(BH$_4$)$_3$(THF)$_n$] (Ln = Sc (n = 2); Ln = La, Nd, Lu (n = 3)) zeigte eine Abhängigkeit von den Ionenradien der Metallzentren. Während für die größeren Seltenerdmetalle Lanthan und Neodym die erwarteten Produkte [(DIP$_2$pyr)Ln(BH$_4$)$_2$(THF)$_2$] (Ln = La (**13**), Nd (**14**)) entstanden, fand bei den kleineren Seltenerdmetallen Scandium und Lutetium eine unerwartete Redoxreaktion eines Borhydridanions mit einer der beiden Imin-Funktionen des Liganden statt, die zu den Produkten [{DIP$_2$pyr*-BH$_3$}Ln(BH$_4$)(THF)$_2$] (Ln = Sc (**15**), Lu (**16**)) führte (Schema II).

Desweiteren wurden die beiden Neodymkomplexe **12** und **14** zusammen mit verschiedenen Cokatalysatoren als Ziegler-Natta Systeme für die Polymerisation von 1,3-Butadien zu Poly-*cis*-1,4-butadien eingesetzt. Mit **12** als Katalysator konnten sehr hohe Aktivitäten und gute Selektivitäten beobachtet werden.

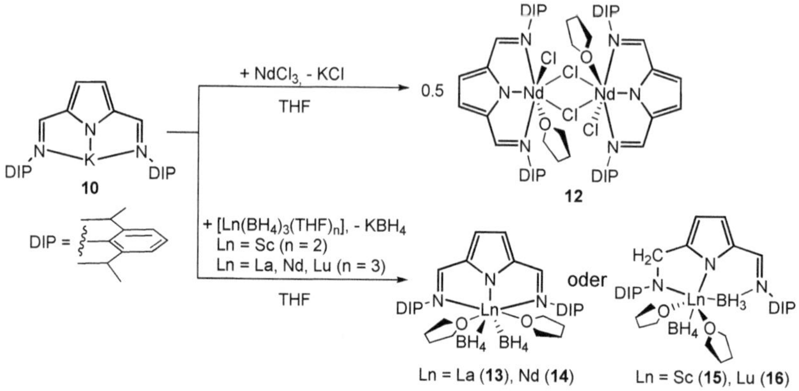

Schema II

Der 2,5-bis{*N*-(2,6-diisopropylphenyl)iminomethyl}pyrrolyl Ligand konnte erfolgreich in die Koordinationschemie der zweiwertigen Lanthanide und der Erdalkalimetalle eingeführt werden. Wie in Schema III dargestellt, erfolgte die Synthese der heteroleptischen Iodokomplexe [(DIP$_2$pyr)LnI(THF)$_3$] (Ln = Sm (**19**), Eu (**20**), Yb (**21**)) sowie [(DIP$_2$pyr)MI(THF)$_n$] (M = Ca (**24**), Sr (**22**) (n = 3); Ba (**23**) (n = 4)) durch Salzmetathesereaktionen von [(DIP$_2$pyr)K] (**10**) mit entweder den wasserfreien Lanthaniddiodiden oder den analogen Erdalkalimetaldiiodiden. Überraschenderweise lagen alle Produkte **19-24**, unabhängig von ihren Ionenradien, im Festkörper als Monomere vor. Die Koordinationssphere des jeweiligen Metallzentrums ist durch die zusätzliche Koordination von THF-Molekülen gesättigt, was ein sehr seltenes Strukturmotiv in der Koordinationschemie der größeren zweiwertigen Lanthanide und der Erdalkalimetalle darstellt. Zusätzlich interessant ist der unterschiedliche Koordinationsmodus des (DIP$_2$pyr)$^-$-Liganden, der in der Festkörperstruktur der Calciumverbindung **24** eine η^2-Koordination aufweist, während die übrigen Komplexe **19-23** η^3-koordiniert vorliegen (Schema III). Dies ist besonders ungewöhnlich, da Calcium und Ytterbium normalerweise dazu neigen, isostrukturelle Verbindungen zu bilden.

Um katalytisch aktive Spezies zu erhalten, wurden die Iodokomplexe [(DIP$_2$pyr)MI(THF)$_3$] (M = Ca (**24**), Sr (**22**)) mit [K{N(SiMe$_3$)$_2$}] umgesetzt (Schema IV) und die so erhaltenen Produkte [(DIP$_2$pyr)M{N(SiMe$_3$)$_2$}(THF)$_2$] (M = Ca (**25**), Sr (**26**)) hinsichtlich ihrer Aktivität in der intramolekularen Hydroaminierung von Aminoalkenen und einem Aminoalkin untersucht. Unglücklicherweise erwies sich der Anwendungsbereich der Katalysatoren **25** und **26** als eingeschränkt, da unerwünschte Nebenreaktionen in Form von Alken-Isomerisierung und Imin-Enamin-Tautomerisierung auftraten. Trotzdem zeigten beide Verbindungen hohe Aktivitäten und Ausbeuten, wobei sich der Calciumkomplex **25** als der reaktivere erwies, der hinsichtlich seiner Aktivität durchaus mit dem β-Diketiminatocalciumkomplex [{(DIPNC(Me))$_2$CH}Ca{N(SiMe$_3$)$_2$}(THF)] vergleichbar ist.

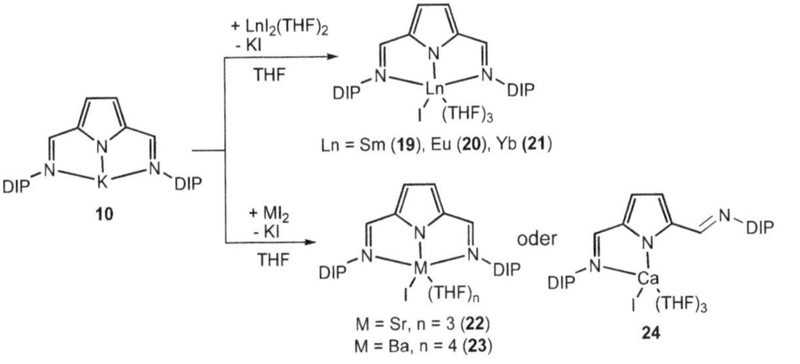

Schema III

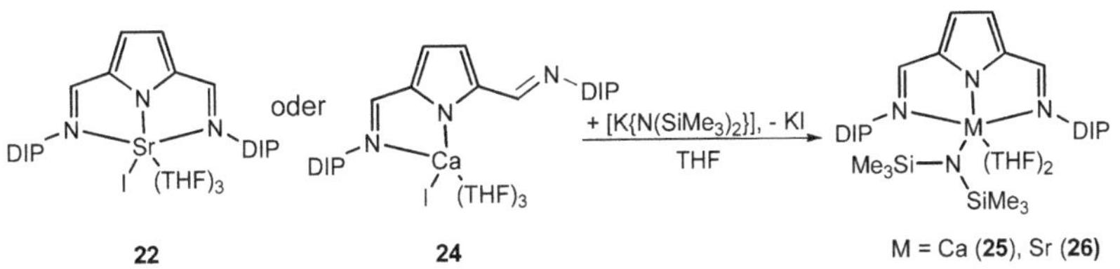

Schema IV

Weiterhin wurden Seltenerdmetallalkylkomplexe zweier Imidazolin-2-imid-Derivate von M. Tamm *et al.* synthetisiert, die im Rahmen der vorliegenden Arbeit als Katalysatoren in Hydroaminierungsreaktionen untersucht wurden. Beide Verbindungen erwiesen sich als katalytisch sehr aktiv und selektiv in der intramolekularen Hydroaminierung von Aminoalkenen und einem Aminoalkin. Obwohl sie nicht mit analogen Metallocenen konkurrieren können, sind sie dennoch neuartige katalytisch aktive Postmetallocene.

6. References

[1] K. H. Wedepohl, *Geochim. Cosmochim. Acta* **1995**, *59*, 1217.

[2] N. Kaltsoyannis, P. Scott, in *The f elements*, Oxford Chemistry Primers, Oxford, **1999**.

[3] E. Riedel, in *Anorganische Chemie, 6. Auflage*, de Gruyter, **2004**.

[4] R. Anwander, *Appl. Homogeneous Catal. Organomet. Compd.* **1996**, *2*, 866.

[5] F. T. Edelmann, *Top. Curr. Chem.* **1996**, *179*, 247.

[6] R. Anwander, *Appl. Homogeneous Catal. Organomet. Compd. (2nd Ed.)* **2002**, *2*, 974.

[7] H. C. Aspinall, *Chem. Rev.* **2002**, *102*, 1807.

[8] M. Shibasaki, N. Yoshikawa, *Chem. Rev.* **2002**, *102*, 2187.

[9] G. Meyer, *Chem. Rev.* **1988**, *88*, 93.

[10] W. J. Evans, *Polyhedron* **1987**, *6*, 803.

[11] W. J. Evans, *J. Organomet. Chem.* **2002**, *647*, 2.

[12] W. J. Evans, *Inorg. Chem.* **2007**, *46*, 3435.

[13] P. B. Hitchcock, M. F. Lappert, L. Maron, A. V. Protchenko, *Angew. Chem.* **2008**, *120*, 1510; *Angew. Chem. Int. Ed.* **2008**, *47*, 1488.

[14] G. Meyer, *Angew. Chem.* **2008**, *120*, 5050; *Angew. Chem. Int. Ed.* **2008**, *47*, 4962.

[15] R. Shannon, *Acta Crystallogr. Sect. A* **1976**, *32*, 751.

[16] G. Wilkinson, J. M. Birmingham, *J. Am. Chem. Soc.* **1954**, *76*, 6210.

[17] J. M. Birmingham, G. Wilkinson, *J. Am. Chem. Soc.* **1956**, *78*, 42.

[18] R. E. Maginn, S. Manastyrskyj, M. Dubeck, *J. Am. Chem. Soc.* **1963**, *85*, 672.

[19] S. Manastyrskyj, R. E. Maginn, M. Dubeck, *Inorg. Chem.* **1963**, *2*, 904.

[20] B. Kanellakopulos, K. W. Bagnall, *MTP (Med. Tech. Publ. Co.) Int. Rev. Sci.: Inorg. Chem., Ser. One* **1972**, *7*, 299.

[21] M. Tsutsui, N. M. Ely, *J. Am. Chem. Soc.* **1975**, *97*, 1280.

[22] T. J. Marks, G. W. Grynkewich, *Inorg. Chem.* **1976**, *15*, 1302.

[23] W. J. Evans, in *Adv. Organomet. Chem., Vol. 24*, Academic Press, **1985**, pp. 131.

[24] H. Schumann, J. A. Meese-Marktscheffel, L. Esser, *Chem. Rev.* **1995**, *95*, 865.

[25] S. Arndt, J. Okuda, *Chem. Rev.* **2002**, *102*, 1953.

[26] S. A. Cotton, *Annu. Rep. Prog. Chem., Sect. A: Inorg. Chem.* **2006**, *102*, 308.

[27] S. A. Cotton, *Annu. Rep. Prog. Chem., Sect. A: Inorg. Chem.* **2009**, *105*, 276.

[28] F. T. Edelmann, *Coord. Chem. Rev.* **2009**, *253*, 343.

[29] W. J. Evans, B. L. Davis, *Chem. Rev.* **2002**, *102*, 2119.

[30] F. T. Edelmann, *J. Alloys Compd.* **1994**, *207-208*, 182.

[31] G. A. Molander, J. A. C. Romero, C. P. Corrette, *J. Organomet. Chem.* **2002**, *647*, 225.

[32] H. Yasuda, *J. Organomet. Chem.* **2002**, *647*, 128.

[33] S. Hong, T. J. Marks, *Acc. Chem. Res.* **2004**, *37*, 673.

[34] F. T. Edelmann, *Angew. Chem.* **1995**, *107*, 2647; *Angew. Chem. Int. Ed.* **1995**, *34*, 2466.

[35] F. T. Edelmann, D. M. M. Freckmann, H. Schumann, *Chem. Rev.* **2002**, *102*, 1851.

[36] W. E. Piers, D. J. H. Emslie, *Coord. Chem. Rev.* **2002**, *233-234*, 131.

[37] K. Dehnicke, A. Greiner, *Angew. Chem.* **2003**, *115*, 1378; *Angew. Chem. Int. Ed.* **2003**, *42*, 1340.

[38] N. N. Greenwood, A. Earnshaw, in *Chemistry of the elements*, Pergamon Press, Oxford, **1984**, p. 122 and 1430.

[39] S. Harder, *Angew. Chem.* **2004**, *116*, 2768; *Angew. Chem. Int. Ed.* **2004**, *43*, 2714.

[40] F. Weber, M. Schultz, C. D. Sofield, R. A. Andersen, *Organometallics* **2002**, *21*, 3139.

[41] H. Schumann, S. Schutte, H.-J. Kroth, D. Lentz, *Angew. Chem.* **2004**, *116*, 6335; *Angew. Chem. Int. Ed.* **2004**, *43*, 6208.

[42]   S. Datta, M. T. Gamer, P. W. Roesky, *Organometallics* **2008**, *27*, 1207.

[43]   M. Wiecko, S. Marks, T. K. Panda, P. W. Roesky, *Z. Anorg. Allg. Chem.* **2009**, *635*, 931.

[44]   D. J.-J. Brunet, D. D. Neibecker, in *Catalytic Heterofunctionalization* (Ed.: P. D. H. G. Prof. Dr. Antonio Togni), **2001**, pp. 91.

[45]   D. M. Roundhill, *Chem. Rev.* **1992**, *92*, 1.

[46]   M. Johannsen, K. A. Jorgensen, *Chem. Rev.* **1998**, *98*, 1689.

[47]   M. Nobis, B. Drießen-Hölscher, *Angew. Chem.* **2001**, *113*, 4105; *Angew. Chem. Int. Ed.* **2001**, *40*, 3983.

[48]   J. Seayad, A. Tillack, C. G. Hartung, M. Beller, *Adv. Synth. Catal.* **2002**, *344*, 795.

[49]   I. Bytschkov, S. Doye, *Eur. J. Org. Chem.* **2003**, 935.

[50]   F. Pohlki, S. Doye, *Chem. Soc. Rev.* **2003**, *32*, 104.

[51]   P. W. Roesky, T. E. Müller, *Angew. Chem.* **2003**, *115*, 2812; *Angew. Chem. Int. Ed.* **2003**, *42*, 2708.

[52]   F. Alonso, I. P. Beletskaya, M. Yus, *Chem. Rev.* **2004**, *104*, 3079.

[53]   J. F. Hartwig, *Pure Appl. Chem.* **2004**, *76*, 507.

[54]   K. C. Hultzsch, *Adv. Synth. Catal.* **2005**, *347*, 367.

[55]   K. C. Hultzsch, *Org. Biomol. Chem.* **2005**, *3*, 1819.

[56]   R. A. Widenhoefer, X. Han, *Eur. J. Org. Chem.* **2006**, *2006*, 4555.

[57]   I. Aillaud, J. Collin, J. Hannedouche, E. Schulz, *Dalton Trans.* **2007**, 5105.

[58]   P. A. Hunt, *Dalton Trans.* **2007**, 1743.

[59]   T. E. Müller, K. C. Hultzsch, M. Yus, F. Foubelo, M. Tada, *Chem. Rev.* **2008**, *108*, 3795.

[60]   P. W. Roesky, *Angew. Chem.* **2009**, *121*, 4988; *Angew. Chem. Int. Ed.* **2009**, *48*, 4892.

[61]   R. M. Beesley, C. K. Ingold, J. F. Thorpe, *J. Chem. Soc. Trans.* **1915**, *107*, 1080.

[62]   C. K. Ingold, *J. Chem. Soc. Trans.* **1921**, *119*, 305.

[63]   M. E. Jung, J. Gervay, *J. Am. Chem. Soc.* **1991**, *113*, 224.

[64]   M. E. Jung, G. Piizzi, *Chem. Rev.* **2005**, *105*, 1735.

[65]   B. Schlummer, J. F. Hartwig, *Org. Lett.* **2002**, *4*, 1471.

[66]   T. E. Müller, M. Beller, *Chem. Rev.* **1998**, *98*, 675.

[67]   M. S. Hill, *Annu. Rep. Prog. Chem., Sect. A: Inorg. Chem.* **2007**, *103*, 39.

[68]   J.-S. Ryu, G. Y. Li, T. J. Marks, *J. Am. Chem. Soc.* **2003**, *125*, 12584.

[69]   M. R. Gagne, C. L. Stern, T. J. Marks, *J. Am. Chem. Soc.* **1992**, *114*, 275.

[70]   Y. Li, T. J. Marks, *J. Am. Chem. Soc.* **1996**, *118*, 9295.

[71]   M. R. Gagne, T. J. Marks, *J. Am. Chem. Soc.* **1989**, *111*, 4108.

[72]   M. A. Giardello, V. P. Conticello, L. Brard, M. R. Gagne, T. J. Marks, *J. Am. Chem. Soc.* **1994**, *116*, 10241.

[73]   F. T. Edelmann, F. H. Anthony, J. F. Mark, in *Adv. Organomet. Chem., Vol. 57*, Academic Press, **2008**, pp. 183.

[74]   S. Bambirra, A. Meetsma, B. Hessen, *Organometallics* **2006**, *25*, 3454.

[75]   N. Meyer, A. Zulys, P. W. Roesky, *Organometallics* **2006**, *25*, 4179.

[76]   F. Lauterwasser, P. G. Hayes, S. Bräse, W. E. Piers, L. L. Schafer, *Organometallics* **2004**, *23*, 2234.

[77]   T. K. Panda, A. Zulys, M. T. Gamer, P. W. Roesky, *Organometallics* **2005**, *24*, 2197.

[78]   M. Rastätter, A. Zulys, P. W. Roesky, *Chem. Commun.* **2006**, 874.

[79]   M. Rastätter, A. Zulys, P. W. Roesky, *Chem. Eur. J.* **2007**, *13*, 3606.

[80]   S. Hong, S. Tian, M. V. Metz, T. J. Marks, *J. Am. Chem. Soc.* **2003**, *125*, 14768.

[81]   H. Kim, T. Livinghouse, J. H. Shim, S. G. Lee, P. H. Lee, *Adv. Synth. Catal.* **2006**, *348*, 701.

[82]   J. Y. Kim, T. Livinghouse, *Org. Lett.* **2005**, *7*, 4391.

[83]   I. Aillaud, K. Wright, J. Collin, E. Schulz, J.-P. Mazaleyrat, *Tetrahedron: Asymmetry* **2008**, *19*, 82.

[84]   J. Collin, J.-C. Daran, O. Jacquet, E. Schulz, A. Trifonov, *Chem. Eur. J.* **2005**, *11*, 3455.

[85]   Y. K. Kim, T. Livinghouse, *Angew. Chem.* **2002**, *114*, 3797; *Angew. Chem. Int. Ed.* **2002**, *41*, 3645.

[86]   D. Riegert, J. Collin, J.-C. Daran, T. Fillebeen, E. Schulz, D. Lyubov, G. Fukin, A. Trifonov, *Eur. J. Inorg. Chem.* **2007**, 1159.

[87]   I. Aillaud, D. Lyubov, J. Collin, R. Guillot, J. Hannedouche, E. Schulz, A. Trifonov, *Organometallics* **2008**, *27*, 5929.

[88]   I. Aillaud, J. Collin, C. Duhayon, R. Guillot, D. Lyubov, E. Schulz, A. Trifonov, *Chem. Eur. J.* **2008**, *14*, 2189.

[89]   D. Riegert, J. Collin, A. Meddour, E. Schulz, A. Trifonov, *J. Org. Chem.* **2006**, *71*, 2514.

[90]   M. R. Burgstein, H. Berberich, P. W. Roesky, *Organometallics* **1998**, *17*, 1452.

[91]   K. C. Hultzsch, *Org. Biomol. Chem.* **2005**, *3*, 1819.

[92]   R. Heck, E. Schulz, J. Collin, J.-F. Carpentier, *J. Mol. Catal. A: Chem.* **2007**, *268*, 163.

[93]   S. Bambirra, H. Tsurugi, D. v. Leusen, B. Hessen, *Dalton Trans.* **2006**, 1157.

[94]   D. V. Gribkov, K. C. Hultzsch, F. Hampel, *Chem. Eur. J.* **2003**, *9*, 4796.

[95]   D. V. Gribkov, K. C. Hultzsch, F. Hampel, *J. Am. Chem. Soc.* **2006**, *128*, 3748.

[96]   G. Zi, *Dalton Trans.* **2009**, 9101.

[97]   M. R. Crimmin, I. J. Casely, M. S. Hill, *J. Am. Chem. Soc.* **2005**, *127*, 2042.

[98]   A. G. M. Barrett, C. Brinkmann, M. R. Crimmin, M. S. Hill, P. Hunt, P. A. Procopiou, *J. Am. Chem. Soc.* **2009**, *131*, 12906.

[99]   M. R. Crimmin, M. Arrowsmith, A. G. M. Barrett, I. J. Casely, M. S. Hill, P. A. Procopiou, *J. Am. Chem. Soc.* **2009**, *131*, 9670.

[100]   M. Arrowsmith, M. S. Hill, G. Kociok-Köhn, *Organometallics* **2009**, *28*, 1730.

[101]   S. Datta, P. W. Roesky, S. Blechert, *Organometallics* **2007**, *26*, 4392.

[102]   S. Datta, M. T. Gamer, P. W. Roesky, *Dalton Trans.* **2008**, 2839.

[103]   S.-i. Orimo, Y. Nakamori, J. R. Eliseo, A. Zuttel, C. M. Jensen, *Chem. Rev.* **2007**, *107*, 4111.

[104]   E. Zange, *Chem. Ber.* **1960**, *93*, 652.

[105]   U. Mirsaidov, A. Kurbonbekov, *Dokl. Akad. Nauk Tadzh. SSR* **1985**, *28*, 219.

[106]   U. Mirsaidov, G. N. Boiko, A. Kurbonbekov, A. Rakhimova, *Dokl. Akad. Nauk Tadzh. SSR* **1986**, *29*, 608.

[107]   U. Mirsaidov, I. B. Shaimuradov, M. Khikmatov, *Zh. Neorg. Khim.* **1986**, *31*, 1321.

[108]   S. M. Cendrowski-Guillaume, G. Le Gland, M. Nierlich, M. Ephritikhine, *Organometallics* **2000**, *19*, 5654.

[109]   S. M. Cendrowski-Guillaume, G. Le Gland, M. Lance, M. Nierlich, M. Ephritikhine, *C. R. Chim.* **2002**, *5*, 73.

[110]   D. Barbier-Baudry, O. Blacque, A. Hafid, A. Nyassi, H. Sitzmann, M. Visseaux, *Eur. J. Inorg. Chem.* **2000**, 2333.

[111]   S. M. Cendrowski-Guillaume, M. Nierlich, M. Lance, M. Ephritikhine, *Organometallics* **1998**, *17*, 786.

[112]   F. Bonnet, C. D. C. Violante, P. Roussel, A. Mortreux, M. Visseaux, *Chem. Commun.* **2009**, 3380.

[113]   N. Barros, M. Schappacher, P. Dessuge, L. Maron, S. M. Guillaume, *Chem. Eur. J.* **2008**, *14*, 1881.

[114]   M. Visseaux, T. Chenal, P. Roussel, A. Mortreux, *J. Organomet. Chem.* **2006**, *691*, 86.

[115] T. V. Mahrova, G. K. Fukin, A. V. Cherkasov, A. A. Trifonov, N. Ajellal, J.-F. Carpentier, *Inorg. Chem.* **2009**, *48*, 4258.

[116] G. G. Skvortsov, M. V. Yakovenko, P. M. Castro, G. K. Fukin, A. V. Cherkasov, J.-F. Carpentier, A. A. Trifonov, *Eur. J. Inorg. Chem.* **2007**, *2007*, 3260.

[117] F. Yuan, Y. Zhu, L. Xiong, *J. Organomet. Chem.* **2006**, *691*, 3377.

[118] F. Yuan, J. Yang, L. Xiong, *J. Organomet. Chem.* **2006**, *691*, 2534.

[119] F. Bonnet, A. R. Cowley, P. Mountford, *Inorg. Chem.* **2005**, *44*, 9046.

[120] M. Visseaux, M. Mainil, M. Terrier, A. Mortreux, P. Roussel, T. Mathivet, M. Destarac, *Dalton Trans.* **2008**, 4558.

[121] P. Zinck, A. Valente, A. Mortreux, M. Visseaux, *Polymer* **2007**, *48*, 4609.

[122] P. Zinck, M. Visseaux, A. Mortreux, *Z. Anorg. Allg. Chem.* **2006**, *632*, 1943.

[123] F. Bonnet, M. Visseaux, D. Barbier-Baudry, E. Vigier, M. M. Kubicki, *Chem. Eur. J.* **2004**, *10*, 2428.

[124] M. Terrier, M. Visseaux, T. Chenal, A. Mortreux, *J. Polym. Sci., Part A: Polym. Chem.* **2007**, *45*, 2400.

[125] J. Thuilliez, R. Spitz, C. Boisson, *Macromol. Chem. Phys.* **2006**, *207*, 1727.

[126] N. Barros, P. Mountford, S. M. Guillaume, L. Maron, *Chem. Eur. J.* **2008**, *14*, 5507.

[127] I. Palard, M. Schappacher, B. Belloncle, A. Soum, S. M. Guillaume, *Chem. Eur. J.* **2007**, *13*, 1511.

[128] I. Palard, A. Soum, S. M. Guillaume, *Chem. Eur. J.* **2004**, *10*, 4054.

[129] S. M. Guillaume, M. Schappacher, N. M. Scott, R. Kempe, *J. Polym. Sci., Part A: Polym. Chem.* **2007**, *45*, 3611.

[130] S. M. Guillaume, M. Schappacher, A. Soum, *Macromolecules* **2003**, *36*, 54.

[131] I. Palard, A. Soum, S. M. Guillaume, *Macromolecules* **2005**, *38*, 6888.

[132] Y. Nakayama, K. Sasaki, N. Watanabe, Z. Cai, T. Shiono, *Polymer* **2009**, *50*, 4788.

[133] I. Palard, M. Schappacher, A. Soum, S. M. Guillaume, *Polym. Int.* **2006**, *55*, 1132.

[134] Y. Nakayama, S. Okuda, H. Yasuda, T. Shiono, *React. Funct. Polym.* **2007**, *67*, 798.

[135] G. Wu, W. Sun, Z. Shen, *React. Funct. Polym.* **2008**, *68*, 822.

[136] D. Barbier-Baudry, F. Bouyer, A. S. M. Bruno, M. Visseaux, *Appl. Organomet. Chem.* **2006**, *20*, 24.

[137] M. Schappacher, N. Fur, S. M. Guillaume, *Macromolecules* **2007**, *40*, 8887.

[138] A. Recknagel, A. Steiner, M. Noltemeyer, S. Brooker, D. Stalke, F. T. Edelmann, *J. Organomet. Chem.* **1991**, *414*, 327.

[139] T. G. Wetzel, S. Dehnen, P. W. Roesky, *Angew. Chem.* **1999**, *111*, 1155; *Angew. Chem. Int. Ed.* **1999**, *38*, 1086.

[140] U. Reißmann, P. Poremba, M. Noltemeyer, H.-G. Schmidt, F. T. Edelmann, *Inorg. Chim. Acta* **2000**, *303*, 156.

[141] S. Wingerter, M. Pfeiffer, F. Baier, T. Stey, D. Stalke, *Z. Anorg. Allg. Chem.* **2000**, *626*, 1121.

[142] M. T. Gamer, P. W. Roesky, *J. Organomet. Chem.* **2002**, *647*, 123.

[143] M. T. Gamer, G. Canseco-Melchor, P. W. Roesky, *Z. Anorg. Allg. Chem.* **2003**, *629*, 2113.

[144] M. T. Gamer, P. W. Roesky, *Inorg. Chem.* **2004**, *43*, 4903.

[145] P. W. Roesky, M. T. Gamer, N. Marinos, *Chem. Eur. J.* **2004**, *10*, 3537.

[146] T. K. Panda, A. Zulys, M. T. Gamer, P. W. Roesky, *J. Organomet. Chem.* **2005**, *690*, 5078.

[147] P. W. Roesky, *Z. Anorg. Allg. Chem.* **2006**, *632*, 1918.

[148] M. Wiecko, P. W. Roesky, V. V. Burlakov, A. Spannenberg, *Eur. J. Inorg. Chem.* **2007**, 876.

[149] M. T. Gamer, M. Rastätter, P. W. Roesky, A. Steffens, M. Glanz, *Chem. Eur. J.* **2005**, *11*, 3165.

[150] R. G. Cavell, R. P. Kamalesh Babu, K. Aparna, *J. Organomet. Chem.* **2001**, *617-618*, 158.

[151] K. Aparna, M. Ferguson, R. G. Cavell, *J. Am. Chem. Soc.* **2000**, *122*, 726.

[152] T. K. Panda, P. W. Roesky, *Chem. Soc. Rev.* **2009**, *38*, 2782.

[153] M. T. Gamer, P. W. Roesky, *Z. Anorg. Allg. Chem.* **2001**, *627*, 877.

[154] M. T. Gamer, S. Dehnen, P. W. Roesky, *Organometallics* **2001**, *20*, 4230.

[155] M. Rastätter, A. Zulys, P. W. Roesky, *Chem. Commun.* **2006**, 874.

[156] M. Rastätter, A. Zulys, P. W. Roesky, *Chem. Eur. J.* **2007**, *13*, 3606.

[157] K. Mashima, H. Tsurugi, *J. Organomet. Chem.* **2005**, *690*, 4414.

[158] Y. Matsuo, K. Mashima, K. Tani, *Organometallics* **2001**, *20*, 3510.

[159] Y. Matsuo, H. Tsurugi, T. Yamagata, K. Tani, K. Mashima, *Bull. Chem. Soc. Jpn.* **2003**, *76*, 1965.

[160] D. M. Dawson, D. A. Walker, M. Thornton-Pett, M. Bochmann, *Dalton* **2000**, 459.

[161] H. Tsurugi, Y. Matsuo, T. Yamagata, K. Mashima, *Organometallics* **2004**, *23*, 2797.

[162] B. A. Salisbury, J. F. Young, G. P. A. Yap, K. H. Theopold, *Collect. Czech. Chem. Commun.* **2007**, *72*, 637.

[163] N. Meyer, M. Kuzdrowska, P. W. Roesky, *Eur. J. Inorg. Chem.* **2008**, 1475.

[164] N. Meyer, *Dissertation*, Freie Universität Berlin, **2007**.

[165] Y. Ikada, H. Tsuji, *Macromol. Rapid Commun.* **2000**, *21*, 117.

[166] G. Rokicki, *Prog. Polym. Sci.* **2000**, *25*, 259.

[167] K. Sudesh, H. Abe, Y. Doi, *Prog. Polym. Sci.* **2000**, *25*, 1503.

[168] A.-C. Albertsson, I. K. Varma, *Adv. Polym. Sci.* **2002**, *157*, 1.

[169] M. Okada, *Prog. Polym. Sci.* **2002**, *27*, 87.

[170] K. M. Stridsberg, M. Ryner, A.-C. Albertsson, *Adv. Polym. Sci.* **2002**, *157*, 42.

[171] A.-C. Albertsson, I. K. Varma, *Biomacromolecules* **2003**, *4*, 1466.

[172] M. Vert, *Biomacromolecules* **2004**, *6*, 538.

[173] L. S. Nair, C. T. Laurencin, *Prog. Polym. Sci.* **2007**, *32*, 762.

[174] D. M. Lyubov, A. M. Bubnov, G. K. Fukin, F. M. Dolgushin, M. Y. Antipin, O. Pelcé, M. Schappacher, S. M. Guillaume, A. A. Trifonov, *Eur. J. Inorg. Chem.* **2008**, *2008*, 2090.

[175] Z. Hou, Y. Wakatsuki, *Coord. Chem. Rev.* **2002**, *231*, 1.

[176] S. Agarwal, C. Mast, K. Dehnicke, A. Greiner, *Macromol. Rapid Commun.* **2000**, *21*, 195.

[177] J. Jenter, P. W. Roesky, N. Ajellal, S. M. Guillaume, N. Susperregui, L. Maron, *Chem. Eur. J.* **2010**, *16*, accepted.

[178] M. Ephritikhine, *Chem. Rev.* **1997**, *97*, 2193.

[179] M. F. Lappert, A. Singh, J. L. Atwood, W. E. Hunter, *J. Chem. Soc., Chem. Commun.* **1983**, 206.

[180] S. Penczek, M. Cypryk, A. Duda, P. Kubisa, S. Slomkowski, *Prog. Polym. Sci.* **2007**, *32*, 247.

[181] S. Penczek, T. Biela, A. Duda, *Macromol. Rapid Commun.* **2000**, *21*, 941.

[182] J. Jenter, P. W. Roesky, *New J. Chem.* **2010**, DOI:10.1039/B9NJ00651F.

[183] P.-C. Kuo, J.-C. Chang, W.-Y. Lee, H. M. Lee, J.-H. Huang, *J. Organomet. Chem.* **2005**, *690*, 4168.

[184] S. J. Trepanier, S. Wang, *Organometallics* **1993**, *12*, 4207.

[185] G. L. J. v. Vliet, F. J. J. d. Kanter, M. Schakel, G. W. Klumpp, A. L. Spek, M. Lutz, *Chem. Eur. J.* **1999**, *5*, 1091.

[186] R. Taube, G. Sylvester, in *Applied Homogeneous Catalysis with Organometallic Compounds (2nd Ed.)*, VCH, Weinheim, **2002**, pp. 285.

[187] D. J. Wilson, *Polym. Int.* **1996**, *39*, 235.

[188] J. Witte, *Angew. Makromol. Chem.* **1981**, *94*, 119.

[189]   A. Fischbach, R. Anwander, *Adv. Polym. Sci.* **2006**, *204*, 155.

[190]   L. Friebe, O. Nuyken, W. Obrecht, *Adv. Polym. Sci.* **2006**, *204*, 1.

[191]   S. K.-H. Thiele, D. R. Wilson, *Polymer Reviews* **2003**, *43*, 581

[192]   V. Lorenz, H. Gorls, S. K. H. Thiele, J. Scholz, *Organometallics* **2005**, *24*, 797.

[193]   S. Maiwald, R. Taube, H. Hemling, H. Schumann, *J. Organomet. Chem.* **1998**, *552*, 195.

[194]   R. Taube, H. Windisch, S. Maiwald, H. Hemling, H. Schumann, *J. Organomet. Chem.* **1996**, *513*, 49.

[195]   W. Gao, D. Cui, *J. Am. Chem. Soc.* **2008**, *130*, 4984.

[196]   J. Jenter, N. Meyer, P. W. Roesky, S. K.-H. Thiele, G. Eickerling, W. Scherer, *Chem. Eur. J.* **2010**, *16*, accepted.

[197]   P. W. Roesky, *Chem. Ber./Recl.* **1997**, *130*, 859.

[198]   P. W. Roesky, *Inorg. Chem.* **1998**, *37*, 4507.

[199]   P. W. Roesky, M. R. Bürgstein, *Inorg. Chem.* **1999**, *38*, 5629.

[200]   N. Meyer, J. Jenter, P. W. Roesky, G. Eickerling, W. Scherer, *Chem. Commun.* **2009**, 4693.

[201]   E. B. Lobkovskii, S. E. Kravchenko, K. N. Semenenko, *Zh. Strukt. Khim.* **1977**, *18*, 389.

[202]   M. F. Lappert, A. Singh, J. L. Atwood, W. E. Hunter, *J. Chem. Soc., Chem. Commun.* **1983**, 206.

[203]   J. Jenter, R. Köppe, P. W. Roesky, *C. R. Chim.* **2010**, accepted.

[204]   F. Bonnet, M. Visseaux, D. Barbier-Baudry, A. Hafid, E. Vigier, M. M. Kubicki, *Inorg. Chem.* **2004**, *43*, 3682.

[205]   A. Klein, R. W. H. Pohl, *Z. Anorg. Allg. Chem.* **2008**, *634*, 1388.

[206]   G. Heckmann, M. Niemeyer, *J. Am. Chem. Soc.* **2000**, *122*, 4227.

[207]   W. Maudez, M. Meuwly, K. M. Fromm, *Chem. Eur. J.* **2007**, *13*, 8302.

[208]   G. R. Giesbrecht, C. Cui, A. Shafir, J. A. R. Schmidt, J. Arnold, *Organometallics* **2002**, *21*, 3841.

[209]   T. V. Petrovskaya, I. L. Fedyushkin, V. I. Nevodchikov, M. N. Bochkarev, N. V. Borodina, I. L. Eremenko, S. E. Nefedov, *Russ. Chem. Bull.* **1998**, *47*, 2271.

[210]   S. Qayyum, A. Noor, G. Glatz, R. Kempe, *Z. Anorg. Allg. Chem.* **2009**, *635*, 2455.

[211]   W. J. Evans, J. W. Grate, H. W. Choi, I. Bloom, W. E. Hunter, J. L. Atwood, *J. Am. Chem. Soc.* **1985**, *107*, 941.

[212]   N. M. Scott, R. Kempe, *Eur. J. Inorg. Chem.* **2005**, *2005*, 1319.

[213]   Z. Hou, A. Fujita, T. Yoshimura, A. Jesorka, Y. Zhang, H. Yamazaki, Y. Wakatsuki, *Inorg. Chem.* **1996**, *35*, 7190.

[214]   J. S. Thrasher, J. B. Nielsen, S. G. Bott, D. J. McClure, S. A. Morris, J. L. Atwood, *Inorg. Chem.* **1988**, *27*, 570.

[215]   D. Barbier-Baudry, S. Heiner, M. M. Kubicki, E. Vigier, M. Visseaux, A. Hafid, *Organometallics* **2001**, *20*, 4207.

[216]   C. Eaborn, P. B. Hitchcock, K. Izod, Z.-R. Lu, J. D. Smith, *Organometallics* **1996**, *15*, 4783.

[217]   P. B. Hitchcock, A. V. Khvostov, M. F. Lappert, A. V. Protchenko, *Z. Anorg. Allg. Chem.* **2008**, *634*, 1373.

[218]   M. N. Bochkarev, G. V. Khoroshenkov, D. M. Kuzyaev, A. A. Fagin, M. E. Burin, G. K. Fukin, E. V. Baranov, A. A. Maleev, *Inorg. Chim. Acta* **2006**, *359*, 3315.

[219]   T. Gröb, G. Seybert, W. Massa, K. Harms, K. Dehnicke, *Z. Anorg. Allg. Chem.* **2000**, *626*, 1361.

[220]   M. Vestergren, B. Gustafsson, A. Johansson, M. Håkansson, *J. Organomet. Chem.* **2004**, *689*, 1723.

[221]   W. J. Evans, T. S. Gummersheimer, J. W. Ziller, *J. Am. Chem. Soc.* **1995**, *117*, 8999.

[222] T. Gröb, G. Seybert, W. Massa, K. Dehnicke, *Z. Anorg. Allg. Chem.* **1999**, *625*, 1897.
[223] G. H. Maunder, A. Sella, *Polyhedron* **1998**, *17*, 63.
[224] M. H. Chisholm, J. C. Gallucci, G. Yaman, *Dalton Trans.* **2009**, 368.
[225] S. Datta, M. T. Gamer, P. W. Roesky, *Dalton Trans.* **2008**, 2839.
[226] H. Sitzmann, F. Weber, M. D. Walter, G. Wolmershauser, *Organometallics* **2003**, *22*, 1931.
[227] K. F. Tesh, D. J. Burkey, T. P. Hanusa, *J. Am. Chem. Soc.* **1994**, *116*, 2409.
[228] H. M. El-Kaderi, M. J. Heeg, C. H. Winter, *Polyhedron* **2006**, *25*, 224.
[229] M. J. Harvey, T. P. Hanusa, *Organometallics* **2000**, *19*, 1556.
[230] M. Sierka, A. Hogekamp, R. Ahlrichs, *J. Chem. Phys.* **2003**, *118*, 9136.
[231] O. Treutler, R. Ahlrichs, *J. Chem. Phys.* **1995**, *102*, 346.
[232] J. P. Perdew, *Phys. Rev. B* **1986**, *34*, 7406.
[233] J. P. Perdew, *Phys. Rev. B* **1986**, *33*, 8822.
[234] A. D. Becke, *Phys. Rev. A* **1988**, *38*, 3098.
[235] S. H. Vosko, L. Wilk, M. Nusair, *Can. J. Phys.* **1980**, *58*, 1200.
[236] F. Weigend, *Phys. Chem. Chem. Phys.* **2006**, *8*, 1057.
[237] F. Weigend, R. Ahlrichs, *Phys. Chem. Chem. Phys.* **2005**, *7*, 3297.
[238] A. Schäfer, H. Horn, R. Ahlrichs, *J. Chem. Phys.* **1992**, *97*, 2571.
[239] R. Ahlrichs, M. Bär, M. Haser, H. Horn, C. Kolmel, *Chem. Phys. Lett.* **1989**, *162*, 165.
[240] G. Jeske, H. Lauke, H. Mauermann, P. N. Swepston, H. Schumann, T. J. Marks, *J. Am. Chem. Soc.* **1985**, *107*, 8091.
[241] D. J. Burkey, T. P. Hanusa, *Organometallics* **1996**, *15*, 4971.
[242] M. Wiecko, *Dissertation*, Freie Universität Berlin, **2008**.
[243] N. Kuhn, M. Göhner, M. Grathwohl, J. Wiethoff, G. Frenking, Y. Chen, *Z. Anorg. Allg. Chem.* **2003**, *629*, 793.
[244] M. Tamm, S. Beer, E. Herdtweck, *Z. Naturforsch.* **2004**, *59b*, 1497.
[245] S. Beer, Cristian G. Hrib, Peter G. Jones, K. Brandhorst, J. Grunenberg, M. Tamm, *Angew. Chem.* **2007**, *119*, 9047; *Angew. Chem. Int. Ed.* **2007**, *46*, 8890.
[246] M. Tamm, S. Randoll, T. Bannenberg, E. Herdtweck, *Chem. Commun.* **2004**, 876.
[247] A. J. Arduengo, R. Krafczyk, R. Schmutzler, H. A. Craig, J. R. Goerlich, W. J. Marshall, M. Unverzagt, *Tetrahedron* **1999**, *55*, 14523.
[248] A. J. Arduengo, H. Bock, H. Chen, M. Denk, D. A. Dixon, J. C. Green, W. A. Herrmann, N. L. Jones, M. Wagner, R. West, *J. Am. Chem. Soc.* **1994**, *116*, 6641.
[249] M. Tamm, D. Petrovic, S. Randoll, S. Beer, T. Bannenberg, P. G. Jones, J. Grunenberg, *Org. Biomol. Chem.* **2007**, *5*, 523.
[250] S. H. Stelzig, M. Tamm, R. M. Waymouth, *J. Polym. Sci., Part A: Polym. Chem.* **2008**, *46*, 6064.
[251] M. Tamm, S. Randoll, E. Herdtweck, N. Kleigrewe, G. Kehr, G. Erker, B. Rieger, *Dalton Trans.* **2006**, 459.
[252] S. Beer, K. Brandhorst, C. G. Hrib, X. Wu, B. Haberlag, J. Grunenberg, P. G. Jones, M. Tamm, *Organometallics* **2009**, *28*, 1534.
[253] S. Beer, K. Brandhorst, J. Grunenberg, C. G. Hrib, P. G. Jones, M. Tamm, *Org. Lett.* **2008**, *10*, 981.
[254] D. Petrovic, L. M. R. Hill, P. G. Jones, W. B. Tolman, M. Tamm, *Dalton Trans.* **2008**, 887.
[255] D. Petrovic, C. G. Hrib, S. Randoll, P. G. Jones, M. Tamm, *Organometallics* **2008**, *27*, 778.
[256] N. Kuhn, M. Grathwohl, C. Nachtigal, M. Steimann, *Z. Naturforsch.* **2001**, *56b*, 704.
[257] T. K. Panda, A. G. Trambitas, T. Bannenberg, C. G. Hrib, S. Randoll, P. G. Jones, M. Tamm, *Inorg. Chem.* **2009**, *48*, 5462.

[258] T. K. Panda, D. Petrovic, T. Bannenberg, C. G. Hrib, P. G. Jones, M. Tamm, *Inorg. Chim. Acta* **2008**, *361*, 2236.

[259] T. K. Panda, C. G. Hrib, P. G. Jones, J. Jenter, P. W. Roesky, M. Tamm, *Eur. J. Inorg. Chem.* **2008**, *2008*, 4270.

[260] A. G. Trambitas, T. K. Panda, J. Jenter, P. W. Roesky, C. Daniliuc, C. G. Hrib, P. G. Jones, M. Tamm, *Inorg. Chem.* **2010**, *47*, DOI: 10.1021/ic9024052.

[261] B. D. Stubbert, T. J. Marks, *J. Am. Chem. Soc.* **2007**, *129*, 4253.

[262] S. Tian, V. M. Arredondo, C. L. Stern, T. J. Marks, *Organometallics* **1999**, *18*, 2568.

[263] M. D. Taylor, C. P. Carter, *J. Inorg. Nucl. Chem.* **1962**, *24*, 387.

[264] P. L. Watson, T. H. Tulip, I. Willams, in *Synthetic Methods of Organometallic and Inorganic Chemistry, vol. 6* (Eds.: W. A. Hermann, G. Brauer), Thieme, Stuttgart, **1997**, pp. 26.

[265] J. L. Namy, P. Girard, H. B. Kagan, *Nouv. J. Chim.* **1977**, *1*, 5.

[266] P. Girard, J. L. Namy, H. B. Kagan, *J. Am. Chem. Soc.* **1980**, *102*, 2693.

[267] R. Appel, I. Ruppert, *Z. Anorg. Allg. Chem.* **1974**, *406*, 131.

[268] R. Miller, K. Olsson, *Acta Chem. Scand.* **1981**, *B35*, 303.

[269] K. Olsson, P. Pernemalm, *Acta Chem. Scand.* **1979**, *B33*, 125.

[270] P. H. Martinez, K. C. Hultzsch, F. Hampel, *Chem. Commun.* **2006**, 2221.

[271] I. Damljanović, M. Vukićević, R. D. Vukićević, *Monatsh. Chem.* **2006**, *137*, 301.

[272] G. M. Sheldrick, *Acta Crystallogr., Sect. A* **2008**, *64*, 112.

7. Appendix

7.1 Abbreviations

<u>General:</u>

Ar	aryl
ATI	aminotroponiminate
Bu	butyl
nBu	n-butyl
iBu	*iso*-butyl
tBu	*tert*-butyl
calcd.	calculated
Cat.	catalyst
conv.	conversion
d	day(s)
DIP	2,6-Diisopropyl phenyl
(DIP$_2$pyr)	2,5-bis{N-(2,6-diisopropylphenyl)iminomethyl}pyrrolyl
(DIP$_2$pyr)H	2,5-bis{N-(2,6-diisopropylphenyl)iminomethyl}pyrrole
Et	ethyl
h	hour(s)
Im^{iPr}	1,3-diisopropyl-4,5-dimethylimidazolin
Im^{DIP}	1,3-bis(2,6-diisopropylphenyl)-imidazolin
L	ligand
Ln	rare earth metal, -ion
M	metal, -ion
Me	methyl
Mes	Mesityl (2,4,6-trimethyl phenyl)
min	minute(s)
Ph	phenyl
Pr	propyl
^{i}Pr	*iso*-propyl
quant	quantitative
R	organic group
r.t.	room temperature
solv	solvent
t	time

T	temperature
THF	tetrahydrofuran
X	leaving group

<u>Polymerization:</u>

CL	ε-caprolactone
D $(=M_w/M_n)$	molar mass distribution
M_w	weight average molar mass
M_n	number average molar mass
M_w/M_n	molar mass distribution
MMA	methyl methacrylate
MMAO	methylalumoxane
n.d.	not determined
PBR	poly-butadiene rubber
ROP	ring-opening polymerization
SEC	size-exclusion chromatography
T_g	glass temperature
theo	theoretical
TOF	turnover frequency

<u>Spectroscopy:</u>

NMR	nuclear magnetic resonance	m	*meta*
br	broad	o	*ortho*
d	doublet	p	*para*
dd	doublet of doublet	VT	variable temperature
m	multiplet	MS	Mass Spectroscopy
qt	quintet	EI	Electron Ionization
s	singlet	IR	Infrared
sept	septet	m	medium
t	triplet	s	strong
J	coupling constant	vs	very strong
i	*ipso*	w	weak

7.2   Curriculum Vitae

Name	**Jelena Laura Anna Jenter**
Date of Birth	19.06.1981
Place of Birth	Berlin
Parents	Ilona and Werner Jenter
Nationality	German
Marital Status	Married to Michal Wiecko

School Education

1987 - 1993	Ikarus-Grundschule (elementary school), Berlin
1993 - 2000	Beethoven-Gymnasium (high school), Berlin
06/2000	Abitur (leaving examination / A-Level), Beethoven-Gymnasium, Berlin

Tertiary Education

10/2000 - 03/2001	Studies: psychology (Diploma), Freie Universität Berlin
04/2001 - 06/2006	Studies: chemistry (Diploma), Freie Universität Berlin
10/2003	Preliminary Examination in chemistry (Vordiplom)
10/2005	Diploma Examination in chemistry
11/2005 - 06/2006	Diploma thesis (Supervisor: Prof. Dr. H.-U. Reißig) *"Model Reactions targeting the Functionalization of the Pyrrolidine Ring of FR 901483"*, Institute of Organic Chemistry, Freie Universität Berlin
23/06/2006	Diploma in Chemistry
08/2006 - 12/2006	Research assistant of Prof. Dr. T. Linker, Universität Potsdam
01/2007 - 02/2008	PhD thesis (Supervisor: Prof. Dr. P.W. Roesky) Institute of Chemistry and Biochemistry, Freie Universität Berlin
01/2007 - 02/2008	Teaching assistant in undergraduate laboratory courses for inorganic chemistry (anorganisch chemisches Grundpraktikum), Freie Universität Berlin

03/2008 - current	PhD thesis (Supervisor: Prof. Dr. P.W. Roesky) Institute of Inorganic Chemistry, Universität Karlsruhe (TH)
03/2008 - 09/2008	Teaching assistant in undergraduate laboratory courses for inorganic chemistry (anorganisch chemisches Grundpraktikum Teil II, Quantitative Analyse),Universität Karlsruhe (TH)
09/2008 - 10/2008	Teaching of a seminar group (first year students, basics of chemistry) in English, International Department, Universität Karlsruhe
10/2008 - current	Teaching assistant in advanced laboratory courses for inorganic chemistry, Universität Karlsruhe (TH)
12/2009 - 02/2010	Research in the group of Prof. Dr. P. Junk and Prof. Dr. G. Deacon, Monash University, Melbourne, Australia; supported by "Karlsruhe House of Young Scientists"

7.3   Poster presentations

J. Jenter, N. Meyer, P. W. Roesky
Synthese und Analytik neuer Borhydrid Lanthanoid Komplexe
GDCH Jahrestagung, Ulm, **2007**.

J. Jenter, N. Meyer, P. W. Roesky
Unusual Reactivity of Lanthanide Borohydride Complexes
OZOM5 (5th Australian Organometallics Discussion Meeting), Sydney, **2010**.

7.4   Publications

1)   T. K. Panda, C. G. Hrib, P. G. Jones, J. Jenter, P. W. Roesky, M. Tamm
     Rare Earth and Alkaline Earth Metal Complexes with Me$_2$Si-Bridged Cyclopentadienyl-Imidazolin-2-imine Ligands: A New Concept for the Design of Constrained Geometry Catalysts
     Eur. J. Inorg. Chem. **2008**, 4270-4279.

2)   N. Meyer, J. Jenter, P. W. Roesky, G. Eickerling W. Scherer
     Unusual Reactivity of Lanthanide Borohydride Complexes leading to a Borane Complex
     Chem. Commun. **2009**, 4693-4695.

3)   J. Jenter, R. Köppe, P. W. Roesky
     Mixed Borohydride-Chloride Complexes of the Rare-Earth Elements
     C. R. Chim. **2010**, in press.

4)   J. Jenter, P. W. Roesky [*]
     Dimeric Complexes of Lithium and Sodium Forming a Tetrametalla-cyclobuta[1,2:1,4:2,3:3,4]tetracyclopentane Structure
     New J. Chem. **2010**, DOI:10.1039/B9NJ00651F.

5)   J. Jenter, N. Meyer, P. W. Roesky, S. K.-H. Thiele, G. Eickerling, W. Scherer
     Borane and Borohydride Complexes of the Rare-Earth Elements – Synthesis, Structures, and Butadiene Polymerization Catalysis
     Chem. Eur. J. **2010**, *16*, 5472-5480.

6)   G. Trambitas, T. K. Panda, J. Jenter, P. W. Roesky, C. Daniliuc, C. G. Hrib, P. G. Jones, M. Tamm
     Rare Earth Metal Alkyl, Amido and Cyclopentadienyl Complexes Supported by Imidazolin-2-iminato Ligands: Synthesis, Structural Characterisation and Catalytic Application
     Inorg. Chem. **2010**, *47*, 2435-2446.

7)   J. Jenter, P. W. Roesky, N. Ajellal, S. M. Guillaume, N. Susperregui, L. Maron
     Bis(phosphinimino)methanide Borohydride Complexes of the Rare-Earth Elements as Initiator for the Ring-Opening Polymerization of ε-Caprolactone: Combined Experimental and Computational Investigations
     Chem. Eur. J. **2010**, *16*, 4629-4638.

8)   T. Li, J. Jenter, P. W. Roesky
     Rare Earth Metal Post-Metallocene Catalysts with Chelating Amido Ligands
     Struct. Bonding **2010**, in press.

---

[*] „Hot Article" in *New J. Chem.*

Danksagung

Mein ganz besonderer Dank gilt Professor Dr. Peter W. Roesky, der mich in seiner Arbeitsgruppe aufgenommen hat und mit interessanten und kreativen Anregungen die gesamte Zeit hindurch begleitet hat. Seine Ideen, Flexibilität und das stets offene Ohr für jedes Problem trugen im Wesentlichen zum Gelingen dieser Arbeit bei. Desweiteren möchte ich mich bei Prof. Dr. Annie K. Powell für die Anfertigung des Zweitgutachtens bedanken.

Dem "Karlsruhe House of Young Scientists" danke ich für die Finanzierung meines dreimonatigen Forschungsaufenthaltes in der Arbeitsgruppe von Glen Deacon und Peter Junk an der Monash University in Melbourne, Australien.

Insbesondere danke ich Dr. Michal Wiecko, Dr. Tianshu Li, Dr. Francesco Grilli und Paul Benndorf für die kritische Durchsicht meiner Arbeit und für ihre hilfreichen Anregungen.

Sibylle Schneider, Petra Smie (ehemals Hauser) und Dr. Michael Gamer danke ich für das Messen der Kristallstrukturen. Bei Frau Berberich und Anja Lühl bedanke ich mich für das Anfertigen der vielen NMR-Spektren. Frau Kayas und Sibylle Schneider danke ich für ihre mütterliche Fürsorge. Auch bei den restlichen Mitarbeitern der Serviceabteilungen der Universität Karlsruhe (TH) sowie der FU Berlin möchte ich mich bedanken, besonderer Dank gilt an dieser Stelle Stefanie Seelbinder, Gabriele Leichle, Sibylle Böcker, Sabine Lude, Herrn Müller, Herrn Munshi und Herrn Rieß (und Mitarbeitern) für ihre Hilfsbereitschaft und ihr freundliches Wesen.

Herzlicher Dank gilt Larissa Zielke (meiner rechten Hand) und Paul Benndorf, die sich ein Labor mit mir teilen und mir die Arbeit mit Freundlichkeit und viel Witz versüßt haben. Allen Kollegen der Arbeitsgruppe Roesky möchte ich für die angenehme Arbeitsatmosphäre danken, besonderer Dank gilt an dieser Stelle meiner ehemaligen Arbeitskollegin Karolin Löhnwitz, sowie Dominique Thielemann, Magdalena Kuzdrowska und Tianshu Li.

Ganz besonderer Dank gilt meiner Familie, die es mir ermöglicht hat, an diesen Punkt zu gelangen. Meiner Mutter Ilona danke ich für ihre Stärke und ihr selbstloses Wesen. Meinem Vater Werner danke ich dafür, dass er zeitlebens mit seinem unglaublich großen Herzen für mich da war. Ich danke meinen geliebten Brüdern Jerome und Jascha dafür, dass sie so sind wie sie sind. Meiner Tante Heike und meinem Onkel Hermann danke ich dafür, dass sie immer für mich da waren. Meinem Mann, Michal Wiecko, danke ich für seine Fürsorge und Unterstützung, die mich auch durch schwere Zeiten begleitet hat, sowie für seine fachlichen Ratschläge, die zum Gelingen dieser Arbeit beitrugen.